51 Advances in Biochemical Engineering Biotechnology

Managing Editor: A. Fiechter

Biotechnics/ Wastewater

With Contributions by
P. Ghosh, S. Hasegawa, R.Ch. Kuhad,
L.C. Lievense, A.J. McLoughlin,
V. Sahai, A. Singh, K. Shimizu,
K. van 't Riet, U. Wiesmann

With 43 Figures and 17 Tables

Springer-Verlag
Berlin Heidelberg GmbH

ISBN 978-3-662-14921-8   ISBN 978-3-540-48062-4 (eBook)
DOI 10.1007/978-3-540-48062-4

Library of Congress Catalog Card Number 72-152360

Springer-Verlag Berlin Heidelberg 1994
Originally published by Springer-Verlag Berlin Heidelberg New York in 1994
Softcover reprint of the hardcover 1st edition 1994

Typesetting: Macmillan India Ltd., Bangalore-25

SPIN: 10119398   02/3020 - 5 4 3 2 1 0 - Printed on acid-free paper

Attention all "Enzyme Handbook" Users:

A file with the complete volume indexes Vols. 1 through 5 in delimited ASCII format is available for downloading at no charge from the Springer EARN mailbox. Delimited ASCII format can be imported into most databanks.

The file has been compressed using the popular shareware program "PKZIP" (Trademark of PKware INc., PKZIP is available from most BBS and shareware distributors).

This file distributed without any expressed or implied warranty.

To receive this file send an e-mail message to:
SVSERV@DHDSPRI6.BITNET.
The message must be: "GET/ENZHB/ENZ_HB.ZIP".

SPSERV is an automatic data distribution system. It responds to your message. The following commands are available:

HELP	returns a detailed instruction set for the use of SVSERV,
DIR (*name*)	returns a list of files available in the directory "name",
INDEX (*name*)	same as "DIR"
CD <*name*>	changes to directory "name",
SEND <*filename*>	invokes a message with the file "filename"
GET <*filename*>	same as "SEND".

Table of Contents

Controlled Release of Immobilized Cells as a Strategy to Regulate Ecological Competence of Inocula

Aiden J. McLoughlin
Dept. Industrial Microbiology, University College Dublin, Dublin 4, Ireland

The ecological competence (the ability of microbial cells/inocula to compete and survive in nature) of laboratory/bioreactor prepared inocula is paramount to commercial exploitation of biotechnological processes initiated by the addition of microbial cultures to natural habitats. Such processes include waste-treatment, bioremediation, dairy and food, agricultural and environmental systems and are characterized by a general inability to regulate the process environment stringently. Such inocula systems will require, as a first step, an efficient formulation and delivery system, based on microenvironmental control, directed at minimizing the lag period and maximizing competitive advantage to the introduced microorganisms.

The use of polymer gels, for example alginate, to immobilize cells has allowed the development of spatially organized microenvironments with control on the degree of protection afforded, the rate of cell release and the juxtapositioning of cells with nutrients and/or selective agents or chemicals.

The characteristics of the gel and its shape has a major influence on the microniche created. Through the control of the gellation process the rate of diffusion of nutrients and the rate of polymer breakdown (or cell release) can be regulated. Surface area to volume ratio can influence the biomass distribution as can the initial biomass loading. The radial gradient created (or the resulting degree of nutrient limitation) in gel beads can have significant influence on both the distribution and behaviour of the immobilized biomass. Thus the combination of immobilization technology and nutrient limitation has resulted in the creation of microenvironments in both space and time dimensions. The resultant inocula delivery systems improve the resistance of the culture and regulate the release of cells with enhanced resistance to stress which is advantageous when "the window of opportunity" to ensure successful colonization can be restricted.

Advances in Biochemical Engineering/
Biotechnology, Vol. 51
Managing Editor: A. Fiechter
©Springer-Verlag Berlin Heidelberg 1994

1 Introduction

The application of ecological principles will inevitably be necessary for the successful exploitation of some novel biotechnological systems currently characterized by their unstable nature. The limited ability of microbial cells to influence their environment is a major restriction when cultures are exposed to dynamic environmental conditions. Thus an enhancement and understanding of microbial plasticity to interruptions or environmental perturbations is fundamental if those process environments, not amenable to strict control, are to be exploited.

Natural evolutionary pressures and the biotechnologist appear to have developed microorganisms in similar directions each developing the specialist skills or characteristics which in turn affects the general robustness of the cell resulting in the relegation of cell types to narrow niches with ever-decreasing capacities to adapt to a wider range of environments.

Evolution, in nature, of procaryotic systems could not proceed towards the selection of a few 'super-competent generalist organisms' but had to move divergently towards a spread of genetic information among a wide range of more specialist organisms [1]. Thus a high degree of specialization exists in individual organisms with respect to their resistance to a particular variable (e.g. oligotrophic, halophilic, thermophilic, psychrophilic, anaerobic etc.) which results in a characteristic structure for microbial communities in most natural systems.

In practice cultures are generally grown as homogeneous cell suspensions in nutrient rich liquid media under optimum conditions. Natural environments or habitats tend to be more complex with a range of surfaces and interfaces, that influence the microbial population which may be relatively slow growing or dormant, sessile, and exposed to nutrient limitation, often resulting in a heterogeneous distribution of microbial populations throughout the environment.

Considering the energetics of the microbial cell, most optimized commercial treatments or recombinant cultures probably represent an inefficient and undesirable system with the ultimate goal of complete conversion of nutrients into products, often superfluous to the cells requirements. Inherently unstable biotechnological processes will arise when such microbial cells optimized for specialist laboratory or bioreactor conditions are exposed to heterogeneous or open systems containing "fitter" or more robust microbial populations with selective or enhanced ability to respond to dynamic conditions.

2 Ecological Competence and Stability of Commercial Inocula

The ecological competence of artificially prepared inocula is paramount to commercial exploitation of biotechnological processes initiated by adding microbial cultures to natural habitats. These include waste-treatment, bio-remediation, dairy and food, agricultural and environmental systems. Such processes are characterized by an inability to regulate stringently the process environment.

Judging from the inconsistency of, for example, biological control reported in the literature [2–5] there appear to be major problems in the development of the technology especially in relation to the stability of the microbial system employed.

The metabolic activity and adaptability or ecological competence of commercial inoculants appears to be a major limitation. For example, Nesbakken and Broch-Due (6) suggested that the inefficacy of many commercial inoculants, used in the ensilement of forage crops, arose from lack of sufficient numbers of bacteria. Such inoculants are added as freeze-dried powder, or immediately after suspension of freeze-dried cultures in water, resulting in insufficient numbers of metabolically active cells – inoculants cultured overnight before application resulted in improved activity. They also highlight that many commercial inoculants contain cultures that are not adapted to forages.

Commercial microbial inocula will require a high degree of robustness if exposed to diverse environments. Technologies based on the addition of inocula to what are basically natural habitats must consider, for example, that with very few exceptions, these habitats are heterogeneous. This heterogeneity may include discontinuity of flow, gas/liquid/solid interfaces, temperature and pH variations, availability of both simple and complex substrates, multiple nutrient limitations and competition with indigenous microbial populations. Further heterogeneity may arise due to the release of solutes from particulate organic matter or complex polymers which in turn may depend on competing microbial enzyme activities. Such degradation processes can result in chemical and physical gradients leading to a range of microenvironments that ultimately gives rise to transient states in the metabolic functioning of the cells.

Environments may also range from (a) the original ecological niche from which the culture was first isolated, (b) the bioreactor used to grow the inoculum, (c) those conditions pertaining during unit operations associated with harvesting, storage, transport and delivery of the inoculum to its ultimate site of action, (d) competition and predation from indigenous microflora. In addition the exploitation of microbial inocula will depend on commercial product parameters such as the convenience and ease of handling. Ideally it should be possible to bulk produce the inoculum and also handle and store it like a chemical.

Likely successful strategies to overcome the changing or dynamic environmental conditions would appear to be based on some form of spatial organization involving: (a) Protection–through the creation of micro-environments, providing the desired conditions or populations and the physical regulatory systems, to minimize the effect of fluctuations in the macroenvironment. Protection from competition, predation and lysis would also be important. (b) Controlled release – through the creation of niches or microenvironments which assist the cells to adapt to the new environmental conditions and then release the adapted cells under controlled conditions.

3 Motility, Juxtapositioning and Spatial Organization

Studies on microbial ecology have demonstrated that spatial organization is a fundamental process in microbial ecosystems. At the population level cells respond to environmental changes by altering their spatial relationships. Such strategies may be necessary when, for example, nutrients are not completely mixed or distributed throughout the environment. An example would be the addition of rhizobia inocula to soil – in order to proliferate a symbiotic association with plant roots. Thus the initial step in successful population survival must be to find a host root system.

To overcome such nutrient gradients a fundamental property would appear to be the ability to distribute effectively or move from one location to another. However microorganisms have certain limitations based on their method of locomotion and their method of sensing.

It has been observed that, for example, bacterial cells move with intent, accumulating in regions which are of favourable chemical composition or have suitable environmental conditions [7]. However due to their unicellular nature they can simply accumulate (through rapid cell division) by remaining associated with a food source. Lauffenburger et al. [8] considered the exploitation of such chemotactic response in biotechnological processes and its value in the potential control of microbial population dynamics in non-mixed systems.

It is tempting to assume that as animals move in order to hunt for food consequently the objective for microbial movement is similar, however, such a simplistic analysis ignores the rheological properties of liquid phases relative to microbial cells. Movement for various types of organisms can be fundamentally different, for example, an animal can swim by pushing fluid backward against the inertia of the fluid, momentum is conserved which results in forward motion. This reciprocal type motion is not effective when swimming at low Reynolds number (Nre) hence the "flexible oar" or flagella type motion which is associated with bacteria (an analogous low Nre situation would be an animal swimming in mud or molasses). In most natural environments bacterial motility is associated with low Nre, consequently cells cannot easily escape from their environment and are prisoners of the bulk of the fluid phases.

Bacterial motility is influenced by a variety of environmental stimuli. It is significant that often the stimulus is first transduced by the general physiology of the cell rather than detected by a specialised sensory receptor. Thus energy state becomes a basic sensory input. Taking account of the single cell nature of bacteria, the form of growth, and cell energetics it is most unlikely that microbial systems are designed to "hunt" for food, it is more likely that chemotactic responses and motility are designed to prevent cells from becoming dissociated from existing food sources and thus impose a spatial order based on food supply.

Convection will distribute the bacterial population throughout the bulk fluid or macroenvironment much more efficiently than bacterial motility. The impressive relative speed of a bacterial cell (20–30 cell lengths per s) compared with that of fast animals such as the cheetah (4 body lengths per s) suggests a highly mobile organism. However because of its microscopic size it is simply embarking on a microscopic journey despite its impressive relative speed. Effectively, transport of nutrients or wastes is controlled by diffusion. For example, to increase its food supply by 10% the cell would have to move at a speed of 20 times that which has been observed [9]. Simply, microbial cells cannot move fast enough to overcome the speed or rate of diffusion of small molecular weight nutrients [10, 11].

The nature of cell movement probably further indicates its major function as a sensory mechanism. Bacterial cells move in two phases. First the cell moves in a relatively straight pathway which is then altered to the second phase in which the cell tumbles or alternates its orientation randomly to a new direction. Brown and Berg [12] developed model systems to describe the effects of substrate concentration on the first phase which depends on chemical concentration – when high the path length is extended. Thus in a gradient there is a bias resulting in a drift towards the nutrients. It appears that the cell interprets a spatial gradient as a temporal one – the cell moving in the gradient will sense an increase/decrease in nutrient concentration with time. Microbial cells, due to their small size, are unlikely to identify gradients by sensing the concentration along the length or width of the stationary cell. Consequently the cell moves in order to experience or sense the environment.

A number of studies confirm the sensory nature of microbial movement. Adler [13, 14], Adler and Templeton [15] and Adler et al. [16] studied the responses of *Escherichia coli* to spatial gradients and established a quantitative dose-response curve. These studies showed that (1) chemicals that are extensively metabolized need not necessarily attract, (2) chemicals that are not metabolized may attract, (3) transport of a chemical into the cell is neither necessary nor sufficient for it to attract. These results probably reflects the fact that such cells are sensing an environment likely to provide suitable nutrients rather than simply responding to specific chemicals and highlights the need for a holistic approach to understand the ecological significance of cell movement.

Thus motility in the microbial world is probably not primarily associated with dissemination of cells or seeking out new food sources but rather is directed towards a bias in favour of remaining associated with (or "immobilized" within) suitable environments when encountered. The balance between cells becoming dissociated from their food source (due to bulk flow of the liquid phase) and remaining associated with the food source (due to motility or chemotactic response) probably represents a slow release mechanism which allows populations to continuously test their environment and to distribute when conditions are favourable.

The activity, dissemination and distribution of cells throughout an environment probably is more dependent on the improved ecological competence arising from the ability to sense, create and exploit microenvironments than from motility directly. Gannon et al. [17] have examined a number of bacterial strains with respect to their ability to move with water through soil. The presence of flagella did not correlate with transport whereas retention was statistically related to size. This study highlights that a number of interacting characters determine whether organisms are transported. Thus the distribution of cells, arising from inocula, throughout an environment such as soil is probably mainly influenced by the ecological competence of the cell to survive the various phases or environmental conditions encountered.

Movement can thus result in a spatial heterogeneity or juxtapositioning of cells within ecosystems. In natural ecosystems microbial cells often form spatial organized ecosystems that act as protective microniches. These "buffer zones" or microenvironments that interface between the microbial cell and the macro-environment may include surface growth, film formation or floc formation. An essential aspect of these strategies appears to be the fixing or immobilization of a large proportion of the population resulting in the microbial cells being physically confined or localized, often in a polymer matrix, in a certain defined region or space. The influence of such structures on the regulation of diffusion to and from the immobilized cells must be a significant mechanism in protection.

This spatial organization appears to be a natural microbial strategy to overcome changing environments. It is also recognized as a basic ecological principle in both plant and animal population ecology – Allee's principle relating the degree of aggregation and overall densities that result in optimum population growth and survival [18]. For example, clumps of plants or animal herds demonstrate both the competitive and cooperative aspects – too many members exhaust the food resource but an appropriate sized group enhances overall survival.

To date, based on the number of papers published, the major interest in the application of immobilization technology is in the area of process intensification such as the opportunity to convert the traditional batch process to the higher rate continuous one. Consequently advantages cited for immobilized cell versus free cell systems include: (a) reduction in the cost of bioprocessing due to repeated use of biocatalyst, (b) the maintenance of higher cell densities and (c) the provision of a system providing minimal cost for separation.

Relatively less attention has been focussed on the unique effects of immobilization on microbial physiology – often the fact that immobilized cells behaved differently from free cells has been considered to be a disadvantage. An obvious benefit of immobilized cell technology is in the control and exploitation of the unique microenvironment (especially that associated with gels) created in such systems, specifically, the potential of this microenvironment in the stabilization of microbial cultures and enzymes [19].

The high level of process control that may be exerted on such microenvironment offers special advantages. Effectively it means that in the case of gel bead immobilization one has the potential to introduce numerous "microreactors" into the macroenvironment.

4 Application of Spatial Organization and Microenvironments in the Stabilization of Bioprocesses

Two examples of immobilization leading to spatial organization or juxtapositioning relevant in biotechnological processes can be seen in the exploitation of:
— Natural flocculation mechanisms used to create microenvironments resulting in stabilized microbial processes such as those found in wastewater treatment systems.
— Immobilization techniques used to create microenvironments that enhance inoculum viability.

4.1 Flocculation

Single microbial cells, for example, can respond to stress such as nutrient limitation by: (a) movement or motility based on trophism, towards a suitable microenvironment. (b) altering single cell state to that of flocs or aggregates resulting in the formation of microenvironments consisting of consortia of immobilized cells.

Activated sludge flocs (involving mixed populations) and yeast flocculation (usually involving only one yeast type as in brewing) are examples of spatially organized ecosystems that are controlled by process parameters. However many studies have concentrated on the physical aspects such as settling properties, floc strength etc. [20–23] rather than the physiological ones. These studies do, however, give some insight into the ecological significance of flocculation and especially the microenvironments created.

Spatial organization such as flocculation appears to be a response to starvation or physiological state [24, 25, 26, 27] and both enhance food utilization and provide a degree of protection to cells. The juxtapositioning arising from aggregation of discrete microbial cells may result in complex metabolic interactions.

Jones et al. [28] demonstrated model systems of defined immobilized consortia and their ability to carry out the biogenesis of methane. Conrad et al. [29] described a specialized metabolic function for cellular aggregation or floc formation in anaerobic digestion systems. The juxtapositioning of microbial species resulted in a spatial organization of syntrophic metabolism [30, 31].

Floc structure enhances utilization of substrates in nutrient limited conditions. Complex substrates such as proteins are concentrated in the vicinity of the biomass [32]. The enzyme activity required for substrate breakdown is mainly cell bound [33, 34]. This juxtapositioning of substrate, enzyme activity and biomass confers ecological advantage in competition for breakdown products.

Flocculation also provides a more efficient utilization of nutrients in that any product leaked or released from a cell can be utilized by a neighbouring one. Cell processes, for example wall synthesis, can result in significant leakage occurring, indeed it is postulated that 50% of cell wall material can be turned over per generation [35]. Studies on *E. coli* have shown that peptides released from peptidoglycan can be used as precursors to form new wall material [36, 37].

In microbial flocs the leaked products of one type of microorganism may become the nutrients of another, for example, algal cells leak organic matter which can be used as a concentrating mechanism by bacterial populations. Azam and Ammerman [38] have proposed a microbial interaction whereby the chemotactic abilities and properties of bacterial cells allow them to sense and utilize higher nutrient concentration regions around algal cells. Jackson [39] examined the significance of the size of phytoplankton aggregates or flocs in relation to leakage. If leakage occurs from cells at a constant rate then smaller spatial organizations or flocs will have lower concentrations surrounding them compared with larger flocs. As the biomass contained within a spherical floc increases approximately as the square of the radius the concentration of the leaked material will decrease faster with increasing distance from a small floc than from a larger one.

Spatial organization (such as flocculation) resulting in juxtapositioning provides a mechanism directed at closing the system so that material lost from one cell can be efficiently captured by neighbouring ones. Thus material lost due to death and lysis can be readily utilized by neighbouring cells. In natural populations death and growth are complementary processes and there is little doubt that death and lysis are fundamental parts of the process of growth and survival within such microenvironments [40, 41].

Spatial organization based on flocculation or aggregation appears to be a means of protection during adverse environmental conditions. During starvation a fraction of the population in the centre portion of the floc will be protected and survival will be enhanced as some of the cells lyse and release nutrients that are easily recycled. Thus a fraction of the population is sacrificed so that some survive.

Direct observations on the interior of, for example, activated sludge flocs indicate an abundant presence of extracellular polymers in amorphous forms surrounding microorganisms in most of the flocs. Thus polymers form a matrix to connect most of the microorganisms together and to maintain the integrity of the flocs through cell immobilization.

Extracellular polymers whether originating from lytic activity or biological synthesis and excretion have been detected on, for example, activated sludge surfaces [27]. A number of investigators have proposed that activated sludge flocs are formed by flocculation of bacteria with naturally occurring polymers acting as floc agents. Tenney and Stumm [42] proposed that biological flocculation occurs through the formation of bridges between cells where naturally occurring polyelectrolytes serve as the bridging polymer. Busch and Stumm [43] showed that polymers extracted from bacterial culture by centrifugation were capable of interparticle bridging. A role for polymers released from cells through lysis has been suggested by Pavoni et al. [27] and Nishikawa and Kuriyama [44] and the polyelectrolyte nature has been shown to enhance flocculation of both sludge and bacterial cultures [45].

Production of extracellular polymers is a common phenomenon in other ecosystems also, for example, most soil microorganisms produce polymers [46] and a number of functions, relating to ecological competence, have been associated with such polymers. These include adhesion, protection against predation and dessication [47–49]. Martens and Frankenberger [50] suggested that polymers, produced by soil bacteria, may also have a function in enhancing enzymatic stability. Microorganisms may retain their extracellular enzymes within these matrices and possibly protect the proteins from breakdown. Such protection would allow an increased exploitation of nutrients in the soil environment.

One of the disadvantages encountered in studies on spatial organization and microenvironments in natural systems has been the number of interacting variables resulting in difficulty in defining, for example, floc structures [21] and in controlling parameters [51, 52]. However the development of immobilization technology, especially the use of natural polymers in gel formation has allowed some degree of process control over spatial organization.

4.2 Immobilization of Inocula

In developing inocula for industrial, agricultural or environmental systems it is essential, in order to stabilize the process and minimize the lag period, to transfer a viable, competitive microbial culture from the inoculum production unit to the process unit. In industrial systems with a high level of process control, it is possible to achieve this through matching the conditions in each unit and then minimizing the transfer time. However in less-controlled biological systems especially those associated with food, agricultural or environmental processes the aseptic conditions of the pure culture inoculum unit contrasts with the complex interacting microbial populations present in, for example, forages, soils or waste treatment systems. Also the time lag between inoculum production and use might be considerable.

Biological control strategies are dependent on the establishment and maintenance of a threshold level of microbial cells in the environment – any decrease in cell viability may eliminate the beneficial effect. Thus maintenance of ecological competence – the ability of a bacterial cell to compete and survive in nature – is essential. A number of workers have, for example, highlighted that protozoan predation of bacterial cells is a major factor causing the decline of introduced inocula [53–55]. Differences in dynamics, under laboratory conditions, between the same bacterial strain occurring indigenously in soil and inoculated into soil were described [56]. The indigenous strain was more stably maintained which was attributed to the habitable microporous nature of soil and/or physiological properties peculiar to the indigenous strain which appeared to protect against protozoan grazing.

Introduced inocula will require, as a first step, an efficient formulation and delivery system which can offer protection to the microbial culture both during storage and when exposed initially to field conditions. The controlled release of ecologically competent cells will also be a necessary process within such delivery systems. The lack of satisfactory delivery is an obstacle to the advancement of biological control technology [57].

The use of immobilization techniques, based on adsorption, in the preparation of inocula for agricultural purposes has been widely reported. Peat is the most common carrier for commercial inoculants and is generally considered to be the most reliable [58]. Other carriers used include vermiculite (which is a hydrated magnesium aluminium iron silicate, the multilamellate structure appears to provide superior aeration and space for microbial proliferation) [59], mineral soils [60], filter mud [61], expanded clay [62], lignite [63], lignite and stillage [64], charcoal [65], bentonite, bagasse [66].

Effectively many of these porous carriers immobilize cells through entrapment – the cells can diffuse into the porous matrix and after cell growth occurs the cells mobility is impeded by the presence of other cells resulting in entrapment. However it is difficult to achieve as high a cell volumetric packing density as those found in gels.

Further important properties of matrices such as soil relate to the nature of the complex porous aggregates which usually possess a high metal-binding capacity, primarily due to their highly surface active components such as clays, organic materials and hydrous metal oxides, resulting in complex sorption-desorption equilibria between metal cations and soil aggregates [67]. Postma et al. [68,69] have shown that survival of bacterial cells introduced into soils was dependent on their spatial distribution in the soil – small pores possibly act as microniches which, for example, protect from predation by protozoa. Heijnen and van Veen [70] demonstrated that kaolinite or bentonite added to loamy sand improved the survival rate of added rhizobia which they attributed to the creation of protective microhabitats. Capillary pores up to 6 µm in diameter are suggested to be the most favourable microhabitat for bacteria [71] with small pores possibly providing protection from predation by protozoa [69, 70, 72]. Competition from natural microflora will also exert a crucial role in the development of an introduced culture. Bashan [73] investigated the inoculation of soil, supporting wheat, with *Azospirillum brasilense* combined with a number of microbial inhibitors to which *A. brasilense* is resistant. This treatment decreased the rhizosphere population while it increased wheat root colonization by the target microorganism and resulted in improved grain yields. Although the inhibitory substances used in this study were not practical for agricultural use, it highlights the significance of providing competitive advantage to the introduced microorganisms.

Apart from protection the pore diameter of the support is critical with respect to microbial cell loading – Opara and Mann [74] have shown that cell loading of yeasts immobilized on ultraporous brick material is correlated with porosity of the support material. Maximum cell loading will be limited by the available surface area of the support – this obviously increases when porous material is used. A number of studies have examined the correlation of optimum pore diameter with cell loading, which is generally accepted as in the range 4–5 fold the largest diameter of the microorganism but obviously the morphology of the microorganism must also be considered (e.g. single cell reproducing by binary fission or budding or from spores and exhibiting mycelial growth) [75, 76].

Immobilization methods based on entrapping agents such as polymer gels probably allows a greater level of microenvironmental control compared with other immobilization techniques. Klein and Ziehr [77] evaluated the merits of entrapment and adsorption relative to bioreactor operation. Entrapment allows higher biomass loading, biomass retainment and operational stability.

Dommergues et al. [78] first proposed polymer entrapment for Rhizobia as an alternative to carriers such as peat. Jung et al. [79] proposed biopolymers such as alginate as more appropriate. Mugnier and Jung [80] outlined the beneficial properties of polymers which can extend to provision of protection during inoculum manufacture, for example, by minimizing heat transfer when spray-drying. These natural polymers give pseudoplastic characteristics to inocula which immediately recover their viscosity when applied to soils. Also the microbial cells are protected in a microenvironment until the polymer structure has been degraded giving controlled release of the inoculum.

The application of immobilization techniques (based on polymer beads) to inoculants was investigated, using single cells/spores or mycelial forms, by a number of workers who have reported enhanced performance when inocula were immobilized in polymer gels. These used include; mycoherbicides [81], biocontrol agents [57, 67, 82–86, 88–90] Mycorrhizae [91–96] Nitrogen fixation [78, 79, 97, 98], and lactic acid bacteria [99–101]. See Table 1 for the range of microorganisms immobilized.

The protective microenvironment created on immobilization leads to, for example, in the case of lactic acid bacteria [100, 101] after lyophilization and rehydration of the culture, enhanced resistance to pH variations in the environ-

Table 1. Microbial cultures demonstrating enhanced activity when immobilized in polymer beads

Microorganism	Reference
Bacteria - Single Cells	
(a) Symbiotic nitrogen fixation:	
Rhizobium sp.	78, 79
(b) Non-symbiotic nitrogen fixation/rhizosphere bacteria:	
Azospirillum lipoferum	98
Azospirillum brasilense	
Pseudomonas sp.	97
(c) Lactic acid bacteria:	
Lactobacillus plantarum	
Pedicoccus pentosaceus	99, 100, 101
Fungi - Mycelial Forms	
(a) Mycoherbicides:	
Alternaria cassiae	
A. macrospora	
Fusarium lateritium	
Colletotrichum malvarum	
Phyllostica sp.	81
(b) Mycorrhizae:	
Hebeloma cylindrosporum	92
Hebeloma crustuliniforme	91
Laccaria laccata	
Suillus luteus	94
Glomus spp.	96
Glomus sp.	95
Laccaria laccata	
Pisolthus tinctorius	
Descolea maculata	
Elaphomyces sp.	
Setchelliogaster sp.	102
Glomus intraradix	103
(c) Biocontrol agents:	
Trichoderma spp.	
Gliocladium spp.	57, 86
Gliocladium virens	89
T. harzianum	83, 84
Talaromyces flavus	90
Laetisaria arvalis	87
T. flavus	
G. virens	
T. viride	
Penicillium oxalicum	82
Beauveria bassiana	85

ment (Fig. 1a, b) although the degree is influenced by the nature of the protectant used which probably reflects the complex interrelationships which may be established within such microenvironments. The role of both the gel and the protectant is possibly significant in minimizing injury during rehydration of preserved cultures. Improved rates of microbial activity (Fig. 2) leading to an overall enhanced treatment was also observed when the immobilized culture

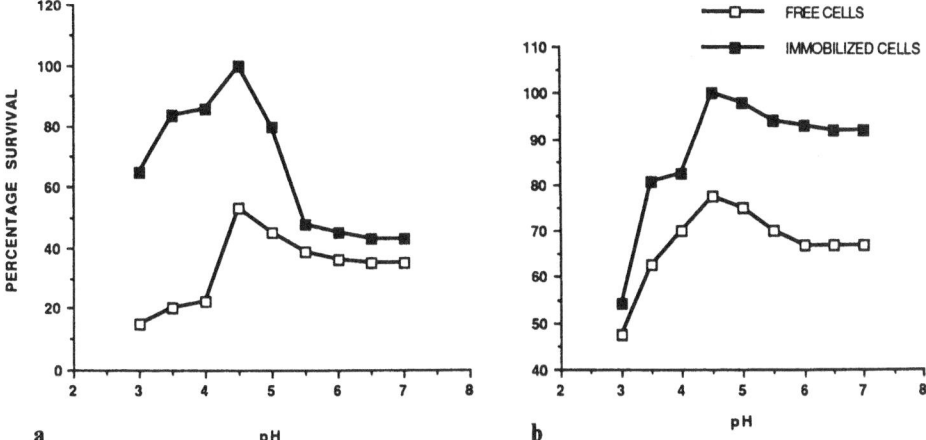

Fig. 1. a. Rehydration of free and immobilized *L. plantarum* cultures with water at different pH values after lyophilization with 1 M glycerol and skim milk as a cryoprotectant. **b.** Rehydration of free and immobilized *L. plantarum* cultures with water at different pH values after lyophilization with 0.75 M adonitol and skim milk as a cryoprotectant

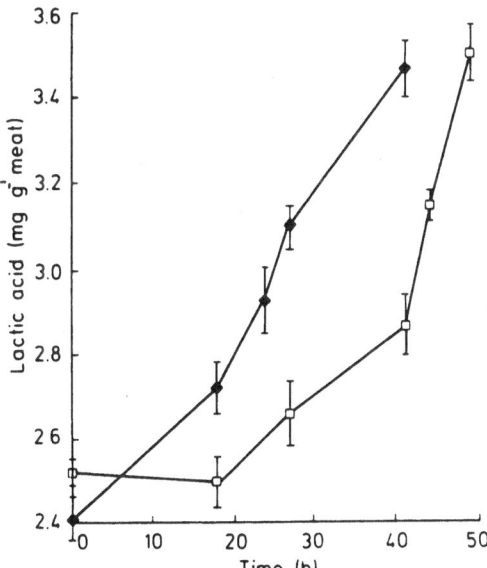

Fig. 2. Lactic acid production during meat processing with free (□) and immobilized (◆) *L. plantarum*

was transferred to a heterogeneous environment and compared with inocula containing similar numbers of free cells.

The enhanced stability (for example plasmid stability) arising on immobilization of cells in naturally occurring polymers has been shown to be influenced by the nature of the matrices, matrix concentration, gel bead volume, biomass loading and environmental conditions such as nutritional limitation, temperature, pH and oxygen have been implicated in the improved stability of some microbial characteristics [104].

Stabilization may arise as a result of: (a) the protective effect conferred by encapsulation in gels, (b) the manipulation of environmental factors such as controlled diffusion rates of nutrients, (c) the ability to incorporate different phases (i.e. solids such as bentonite or peat), food sources, selective chemicals (to inhibit selective fractions of the microflora e.g. predators such as protozoa)

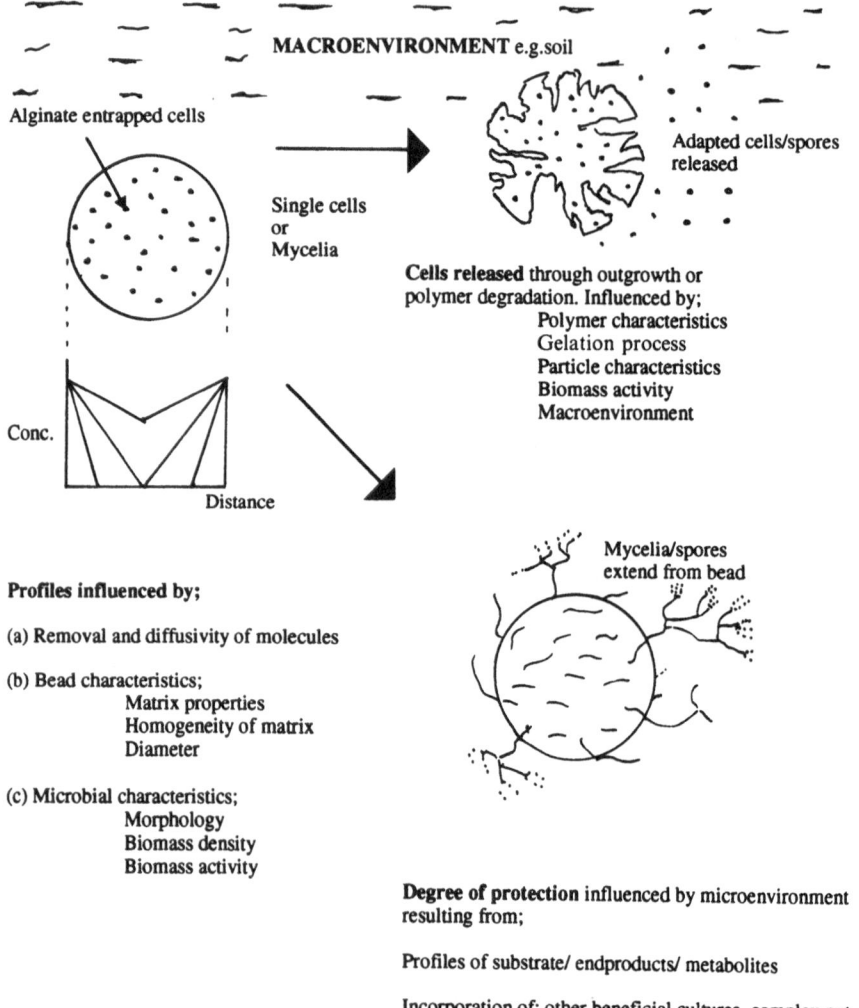

Fig. 3. The establishment of microenvironment and the controlled release of cultures with enhanced ecological competence

and/or protectants into the gel microenvironment [97, 99–101], (d) the ability to direct the release of the cells in a defined and controlled pattern.

Combinations of the benefits of adsorption and entrapment methods have also been investigated. For example, Mauperin et al. (1991) incorporated either sphagnum peat or bentonite in the alginate gel in order to influence the water content of the bead and consequently improve survival of the immobilized cells. Incorporation of peat into the alginate beads resulted in greater levels of hyphal growth. Possible explanations include effects based on the buffering capacity of peat or the possible supply of nutrients or growth stimulants from the peat [105–107].

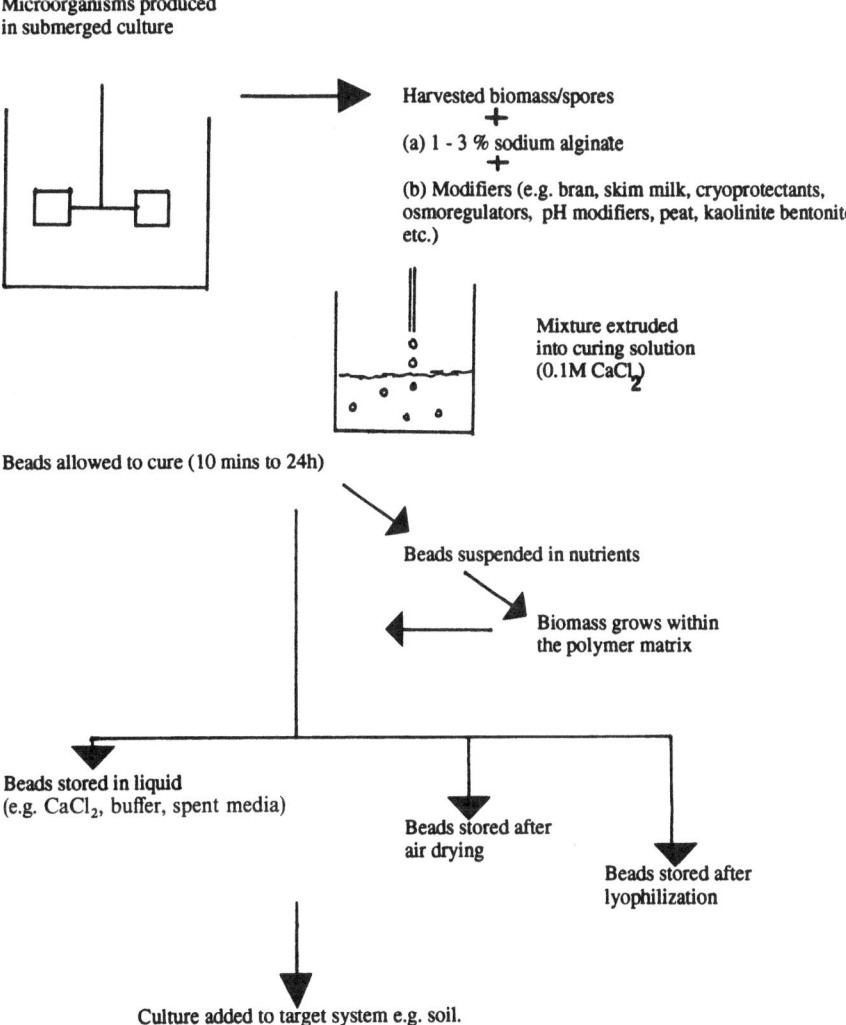

Fig. 4. Protocol for production of inoculum immobilized in polymer beads

Figure 3 outlines the main mechanisms involved in such controlled release systems based on a combination of nutrient limitation and immobilization technology using naturally occurring gels. The physical properties of the gel allows the inoculum to be contained within a protected microniche when applied to the macroenvironment. The particular strategy allows the cells to gradually adapt to the environmental conditions within the macroenvironment. Figure 4 outlines the basic protocol in producing such inocula.

A number of patents have been described, that exploit the potential of gels for controlled release. Mitchell [108] described a system of slow release of fertilizers to soil using alginate gels. Scher [109] described the release of juvenile hormones into aqueous environments from discs of gels as insect control systems. A number of patents have described systems for the controlled release of bioactive materials from alginate gel beads when applied to terrestrial or aqueous environments [110–113].

5 Immobilization and Release of Polymer Entrapped Cells

The further exploitation of spatial organization based on polymer gels will require an understanding of: (1) The unique properties of gels and the influence of the gelation process on these properties, in addition identification of factors influencing the mobility and release of cells will be important, (2) The physical microenvironment created within the gel, the stability of the gel and microenvironment. The specific type of microenvironment encountered in, for example, gel beads especially in relation to effects exerted through biomass distribution and its influence on mass transfer and nutrient limitations. (3) The effect of gel immobilization on microbial physiology and activity.

Unlike natural spatial ecosystems based on extracellular polymers, such as flocs, immobilization technology allows spatial organization to be regulated and controlled much more readily, especially at the level of directing the rate of diffusing nutrients and the controlled release of cells. Woodward [114] outlines the methods of immobilization which include flocculation, adsorption, cell entrapment and metal-link/chelation processes. Immobilization by cell entrapment in polymers is probably the method that gives the greatest level of microenvironmental control. It involves the "trapping" of microbial cells in the pores or interstices of naturally occurring gel polymers such as alginate or carrageenan or within synthetic gels such as acrylamide. Naturally occurring gels and their unique properties would appear to offer the simplest, easiest and safest method to ensure optimum microenvironments for biological activity. However to exploit the full potential of gels it is important to identify the influence of process factors on both gel properties and the microenvironment created within the gel.

5.1 Gel Properties

Gels have both properties of liquid and solids, for example, alginate gels may be formed from 1% alginate and 99% liquid or water yet show solid characteristics such as shape retention and resistance to mechanical stress. A gel consists of physically immobilized water similar to a semipermeable membrane through which low molecular weight, water soluble molecules can diffuse. Water can move into or out of the gel depending on the external environment.

Different polymers, techniques, and gelation processes give a variety of gel characteristics that can influence the type and stability of the microenvironment created.

One of the most widespread natural polymers used in gel formation is alginate. Chemically the alginates consist of linear polymers of 1,4-linked beta-D-mannuronic acid and 1,4-linked alpha-L-guluronic acid arranged in three types of groupings: blocks of mannuronic and blocks of guluronic acid and lastly blocks of alternating mannuronic and guluronic residues. How these blocks are arranged and the proportion of each control the gelling properties of the alginate. For example, gels made with mannuronic acid-rich algin are more deformable, those made with guluronic acid-rich algin are more brittle. The gel is probably a three dimensional network of long chain polymers held together by junction zones and holding water or fluid in the interstices. The junction zones in the case of calcium alginate are probably alpha-L-guluronic blocks from two or more molecules lying parallel and stabilized by calcium ions. Thus the gelling properties and water retention properties of alginate depends on its monomeric composition, sequential arrangements and the lengths of the guluronic blocks in addition to the gelation conditions.

The physical integrity (e.g. size, shape, shape retention etc.) of the microenvironment formed within a gel bead will be determined to a large extent by the gelation process.

5.2 Gelation Conditions

The major factors influencing the gellation of polymer beads are outlined in Table 2. Polymer concentration influences the properties of the gel bead, for

Table 2. Summary of the major factors influencing gellation of polymer beads

1. Chemical properties and concentration of polymer
2. Type and concentration of counterion used
3. Bead diameter and diffusion rate of the counterion
4. Duration and temperature of curing process
5. Process conditions:
   (a) Sterilization
   (b) Addition of nutrients/protectants etc.
   (c) Influence of interfering ions/chemicals
      e.g. carry-over of nutrients, surfactants etc.

example, alginate concentration can influence particle diameter, mechanical strength such as resistance to deformation or compression, as measured by critical compression force (CCF), and permeability of the gel bead formed (Fig. 5).

Gilson et al. [115] developed a diffusional model or equation to predict the time for complete gelling of alginate beads. They have reported the effects of bead diameter, alginate concentration, temperature, calcium chloride concentration and the presence of immobilized cells on gel formation. Gel formation and curing is a dynamic process involving the cooperative binding of counter-

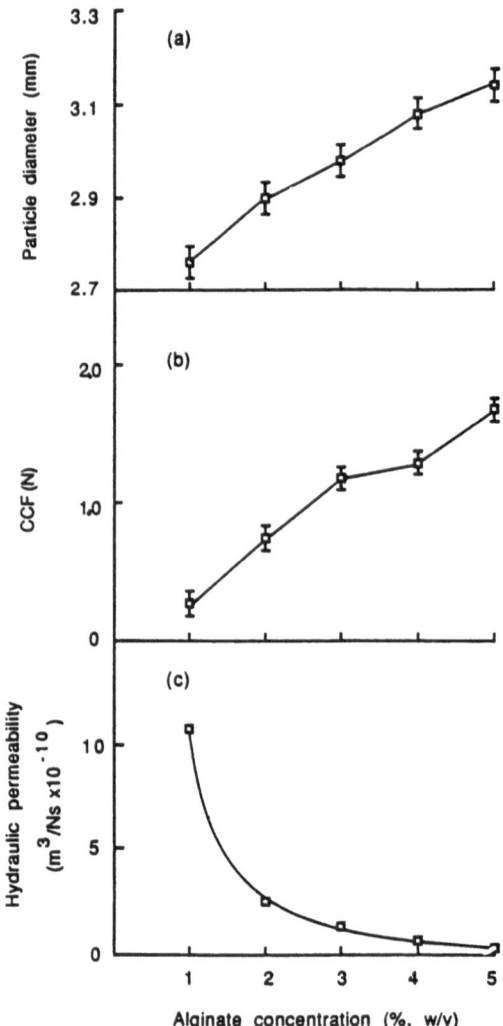

Fig. 5a–c. Effect of alginate concentration of the properties of the gel. **(a)** particle diameter, **(b)** mechanical strength and **(c)** gel permeability

ions to the polymer molecules. The pattern of cross-linking has been termed the "egg-box" model [116]. Thus, for example, alginate bead formation is initiated instantaneously at the bead surface leading to the formation of a membrane of low porosity surrounding each particle [117]. Diffusion of counterions towards the centre of the bead results in ion binding and gel formation in successively deeper layers. The rate at which the gel forms is determined by the diffusivity of the counterion through the preformed outer gel layer [118]. Higher counterion concentrations lead to a compact gel surface with low porosity and very slow gel formation at the particle core [119]. The resulting gel can be nonhomogeneous.

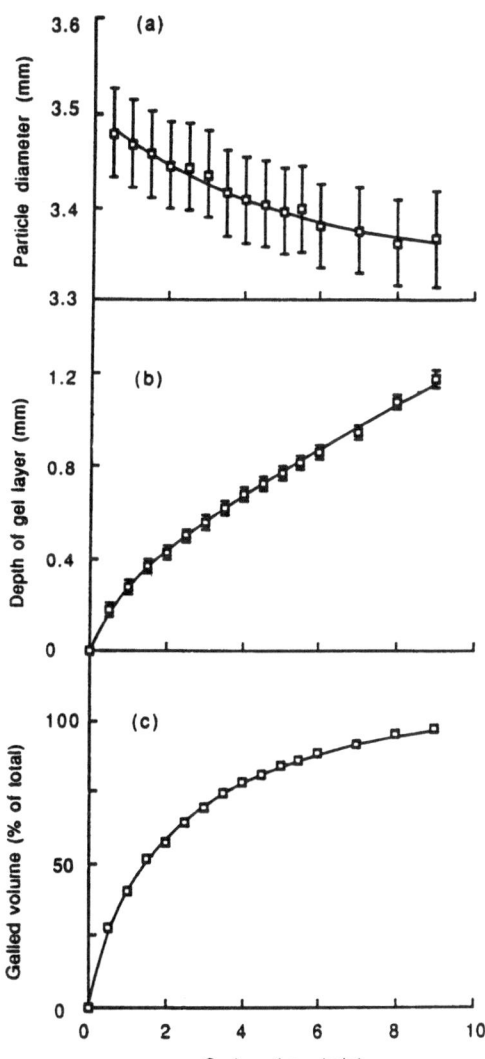

Fig. 6a–c. Physical aspects of gel formation during the initial minutes of curing of standard gel particles in 0.1 M calcium chloride

Figure 6 outlines some of the physical aspects of gel formation during the initial minutes of curing and shows that the depth of the gel volume increases linearly with time between 2 and 10 min. For example Vidoli et al. [120] used very short curing times (8 to 10 min) to encapsulate protoplasts of *Saccharomyces cerevisiae*. The method resulted in a viscous solution of alginate allowing the protoplasts to be free-floating within s polymerized shell. Lower concentrations of counterions (0.05 to 0.1 M) and longer curing times are more desirable in order to achieve relatively homogeneous gels. The duration of the curing time can influence the physical properties of alginate gel particles such as gel bead diameter and mechanical strength (Fig. 7).

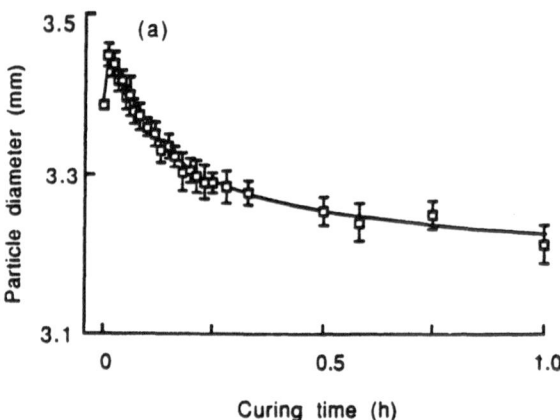

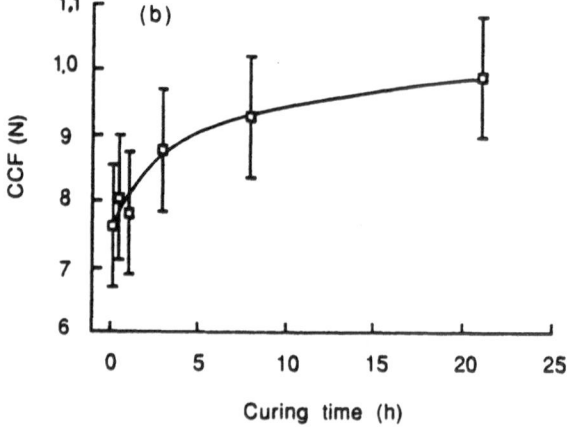

Fig. 7a, b. Effect of curing time on the physical properties of alginate gel particles: (a) diameter, and (b) mechanical strength of standard gel particles cured in 0.1 M $CaCl_2$

Table 3. Effect of the type of counterion on the mechanical properties of 3% (w/v) alginate gel beads

Gel property	Counterion type		
	Calcium	Strontium	Barium
Particle diameter (mm)	2.99 ± 0.05	2.83 ± 0.05	2.75 ± 0.05
CCF (N)	0.76 ± 0.23	1.26 ± 0.23	2.20 ± 0.20
Hydraulic permeability $(m^3/Ns \times 10^{-10})$	2.62	1.70	2.20

A major influence on the properties of the gel formed is the nature and concentration of counterions employed and of other salts to which the gel is subsequently exposed. The choice of counterion may be dictated by its efficiency in forming a gel and its compatibility with the microbial cells. The type of counterion can influence both mechanical strength and permeability with barium giving a stronger gel but less permeable compared with calcium (Table 3). The data of Paul and Vignais [121], Dainty et al. [122] and Tanaka and Irie [123] indicate Ca, Sr and Ba as the counterions of choice, particularly with regard to their resistance to chelating agents. Trivalent ions such as Al and Fe may also be used [124].

The use of multiple counterions has been investigated. Rochefort et al. [119] used $Al_2(NO_3)_3$ treatment after $CaCl_2$ curing. Gray and Dowsett [125] claimed a substantial reduction in gel porosity when a Ca–Zn combination was used but their data are open to interpretations other than gel modification (for example, adsorption phenomena). The operational complexity linked to the apparent minor benefits favours the use of single counterions [121].

The presence of other ions can influence aspects of gel formation. Synersis of gels (free water flowing from the gel) can occur due to the reduced water binding capacity of calcium alginate and/or gel contraction. To overcome this effect the ratio of calcium and/or hydrogen ion content to sodium ions should be reduced [126]. Phosphate ions have been reported to destabilize alginate. However the effect is complex, phosphate concentration may exert different effects depending on its interactions with other ions, cells and the type of alginate. Phosphate promotes gel swelling by chelating calcium in the gel. However at higher concentrations a different phenomenon – gel shrinkage – has been observed [122, 127]. This has been termed "dissolution" or contraction and could be due to a balance between osmotic pressure in the gel and the hydrostatic pressure of the bulk solution. The tendency of gel particles to dissolve at a given phosphate concentration (calcium:phosphate) could be modified by the addition of potassium chloride. Dainty et al. [122] showed that gel particles were less resistant to phosphate in a growth medium containing

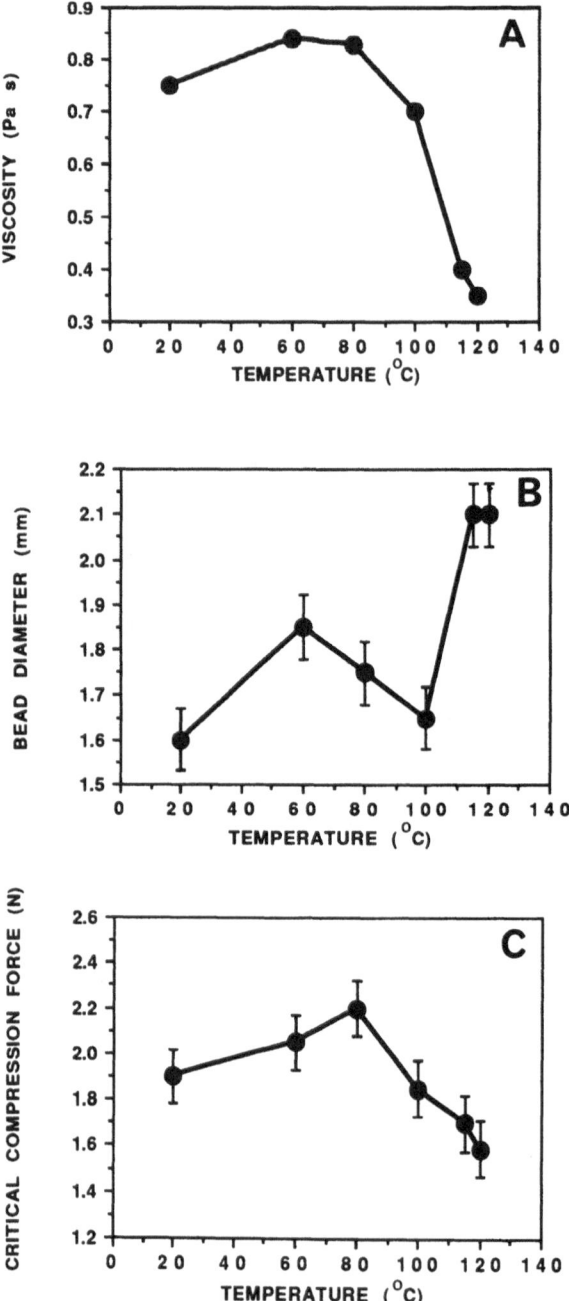

Fig. 8A–C. A) Effect of temperature (15-min exposure time) on the viscosity of 3% (w/v) sodium alginate solutions. **B)** Effect of temperature (15-min exposure time) on the diameter of calcium alginate gel beads (prepared from 3% (w/v) sodium alginate solution). **C)** Effect of temperature (15-min exposure time) on the mechanical strength of calcium alginate gel beads (prepared from 3% (w/v) sodium alginate solution)

Table 4. Physical properties of Ca-alginate beads incorporating 1 M glycerol and skim milk or 0.75 M adonitol and skim milk before and after lyophilization and rehydration

Cryoprotectant	Before lyophilization		
	Shape	Diameter (mm)	Strength (N mm^{-2})
1 M glycerol + skim milk	Spherical	2.09 (± 0.31)[a]	0.38 (± 0.05)
0.75 M adonitol + skim milk	Spherical	2.43 (± 0.27)	0.24 (± 0.04)

Cryoprotectant	After lyophilization and rehydration		
	Shape	Diameter (mm)	Strength (N mm^{-2})
1 M glycerol + skim milk	Spherical	1.19 (± 0.19)	2.04 (± 0.04)
0.75 M adonitol + skim milk	Irregular	1.75 (± 0.71)	0.63 (± 0.02)

[a] Numerical results are given ± standard deviations

sodium and magnesium ions and citrate than in pure phosphate solution. Thus alginate gels are susceptible to chelating agents and calcium substitution by monovalent ions.

In the production of biocatalysts it is usually necessary to employ aseptic techniques. Process conditions such as sterilization procedures can influence gelation. Different heat treatments caused a significant decrease in the degree of polymerization of alginate and consequently modified the gel structure with consequences for mass transfer and cell growth. Other sterilization treatments such as irradiation and ethylene oxide also altered the degree of polymerization of alginate [128]. The duration of heat treatment also influences the characteristics of the gel and immobilized particles produced (Fig. 8a, b, c). Kearney et al. [100] has shown that process factors such as the incorporation of cryoprotectants (glycerol, adonitol etc.) and drying or lyophilization – a step used to enhance viability of bacterial inocula during prolonged storage – can have significant effects on the stability of the gel formed (Table 4). Alginate may be brittle when dry but may be "plasticized" somewhat by the inclusion of glycerol [126].

The influence, on both cell viability and gel stability, of nutrient carry-over when suspensions of the original culture rather than washed cultures are used and the composition of any diluent used [98] must also be assessed.

The influence of surfactants (often added to enhance suspension of spores) has been shown to exert minor effects on the characteristics of alginate beads (Fig. 9). Surfactants may also be added to aid the penetration of curing solution through sodium alginate [129].

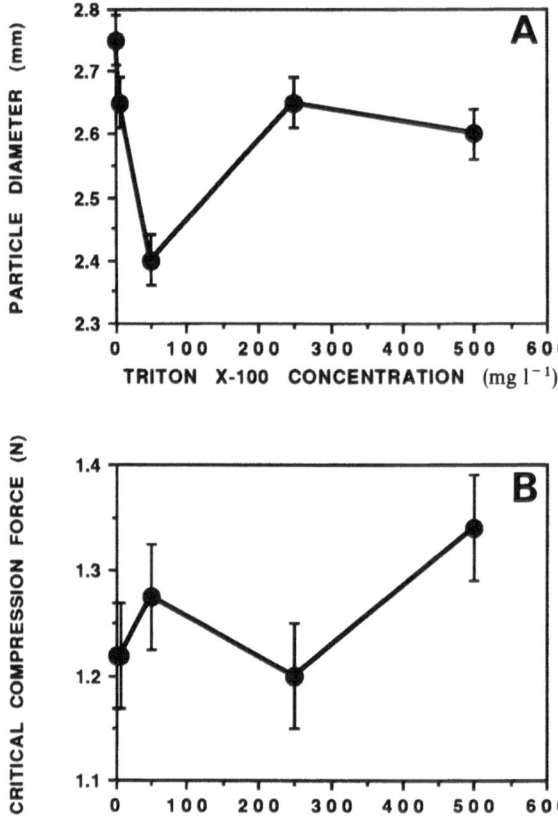

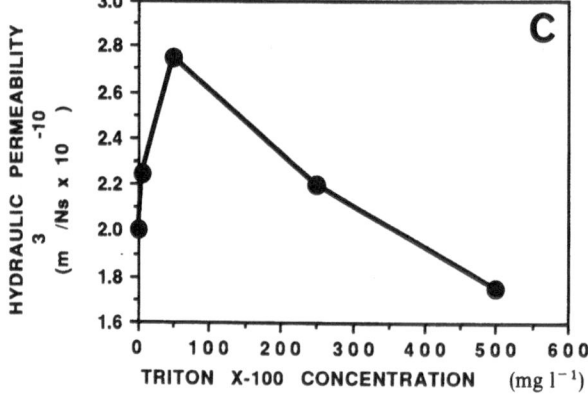

Fig. 9a–c. Effect of addition of surfactant to alginate solution on the physical properties of the resulting gels: (a) diameter and (b) mechanical strength of gel particles, and (c) gel permeability

The adverse effect that gelation conditions may have on viability of the immobilized biomass must also be considered. There are benefits to be gained from modifying curing solutions in relation to both gel stability and biological activity. A number of authors have incorporated buffers in the gelation solution to suit gel formation [121, 130, 131]. Reports directed at modification of the curing solution to suit the physiological characteristics of the immobilized biomass are rarer. Glucose has been added during the formation of gel particles containing *Saccharomyces* spp. [132, 133]. Aeration of the curing solution was reported for the immobilization of *Tagetes* sp. [134]. The loss due to the leaching of compounds (e.g. nutrients or protectants), added to the gel bead during curing, must also be evaluated and balanced.

5.3 Gel Stability and Cell Release

The major factors influencing the degree of gel stability or cell release are outlined in Table 5. The basic aspects of the physical microenvironment created on immobilization in alginate are poorly understood. For example, calcium alginate is probably the most widely used matrix of both immobilized plant and microbial cells, yet this matrix has been described as porous structure [117], honeycomb-like structures with pores ranging from 0.1 to 5 μm [135], macroporous with pore diameters approx. 10 μm [136] and solid structure [137]. Klein et al. [138] report values of 7–17 nm. Casson and Emery [139] discuss the artefactual effects that arise in the methodologies available for examination of the structure of alginate and conclude that the alginate matrix is microporous with a general pore size less than 90 nm.

However by varying the kinetics of the gelation process obviously different degrees of anisotropy or heterogeneity can be obtained. As many of these studies were carried out under different conditions and with non-standard alginates it is not surprising that the resultant gels vary to a substantial degree.

Table 5. Major factors influencing gel stability/cell release

---

1. Gel properties;
    (a) Polymer properties, gelation and hardening techniques employed.
    (b) Characteristics and heterogeneity of the gel formed, e.g. shape (surface/volume ratio), strength, pore size etc.
    (c) Influence of drying/lyophilization, rehydration etc.

2. Mechanism and rate of polymer breakdown
   e.g. chelation or counterion exchange, enzymatic breakdown, disruption due to expanding biomass.
3. Nature of cell leakage or cell outgrowth from the gel matrix
   e.g. Single cells/mycelia, influence of biomass loading, growth rate etc.

In general, stronger gels result in lower permeabilities and presumably smaller pore size – thus higher alginate concentrations form stronger gels which in turn restrict mass transfer which has obvious implications for the micro-environment created. Indeed the degree of homogeneity of the matrix is not well defined. Porosity of surface layers is generally lower than in the core of alginate beads [138, 140]. A membrane-like coating or layer is thought to result from contact between the sodium alginate drop and the curing solution (calcium chloride) at the beginning of gel formation [117]. A similar effect was noted by Kuek and Armitage [137] who observed that calcium alginate appeared to be lost via peeling of a thin outer layer.

Polymer concentration gradients have been observed in gel beads using transmission microscopy. These may vary from circa 10% near the interface with the curing solution to 0.2% at the centre of the gel [141]. Leo [142] in studies, employing X-ray microanalysis, on levels of calcium (the cross linking agent) in alginate particles has also shown heterogeneous distributions of calcium within the bead (Table 6).

Martinsen, Skjåk-Braek and Smidsrød [143] have shown that beads with the highest porosity, greatest strength, lowest shrinkage and best stability towards monovalent cations were prepared from alginate with a content of L-guluronic acid higher than 70% and an average length of the G-block higher than 15. They have shown that above a certain threshold concentration the gel strength increases in proportion to the square of the alginate concentration.

The stability of the gel is an important variable as it influences the duration to which cells are exposed to the microenvironment. The rate of breakdown or the degree of resistance of the alginate bead and consequently release of cells can be altered using a variety of methods. Treatment with cross-linking agents such as dialdehydes or diamines can result in greater mechanical strength. A convenient way of regulating leakage is to bound the matrix with a microporous membrane of suitable porosity thus ensuring an interface between the gel matrix and the surrounding environment.

Birbaum et al. [144] described a number of methods used to enhance stability of beads. In one method the preformed alginate beads were treated with polyethyleneimine followed by glutaraldehyde. Another method involved alginate solution treated with carbodiimide and N-hydroxy-succinimide (to pro-

Table 6. Distribution of calcium ions in 3% (w/v) calcium alginate gel particles cured for 16 h in 0.1 M $CaCl_2$

Location	Relative calcium concentration (% of maximum)
600 μm below surface	16
≈ 50 μm below surface	57
In surface membrane	100

duce active esters), added to cells and extruded into calcium chloride solution to form beads which subsequently were cross-linked with polyethyleneimine. A third method involved alginate solution treated with periodate, resulting in the formation of aldehyde groups, added to cells, extruded into calcium chloride solution followed by cross-linking with polyethyleneimine.

Johansen and Flink [145] described a procedure, using nylon mesh, to reinforce gel strength. Kokufuta et al. [127] stabilized gels by reinforcing a network structure of the gel with polyelectrolyte complex consisting of potassium poly(vinyl alcohol) sulphate and trimethylammonium glycol chitosan iodide. Modifications such as polymers with a high molecular weight and a low mannuronic/guluronic acid ratio gelled by contact with strontium ions resulted in beads that had numerous channels and were still mechanically strong [146].

Properties relating to gel stability may have important implications for exploitation of the microenvironment created. The stability of alginate gel is dependent on the retention of the cross-linking ion with the gel and the preservation of junction zones. Several ionic factors in the surrounding environment may destabilize it by inducing the loss of counterions. These are known to include chelating agents and some monovalent ions [147] Wang et al. [148] have shown that process factors such as shear, ionic strength, cell distribution within the bead and initial cell concentration all influence the cell leakage rate. Through manipulation of the gel properties and bead parameters (such as bead diameter, biomass loading etc.) it is possible to control the rate of polymer breakdown and release of cells.

Alginates can also be degraded by a group of enzymes that catalyse the beta-elimination of the 4-O-linked glycosidic bond with formation of unsaturated uronic acid-containing oligosaccharides [149]. These alginate lyases have been reported to occur in marine fungi [150] and a number of bacteria [151–155]. Typically these alginate lyases show a degree of specificity for either the D-mannuronic or the L-guluronic acid residues at the site of cleavage [156] and consequently could be exploited to control the rate of breakdown of the alginate matrix. Furthermore the alginate lyase gene (alxM) has been cloned and expressed in *E. coli* [157] which offers the potential that the immobilized cell can control its own release from the gel.

The "release" or spread of mycelial forms of microorganisms from polymer gels may also depend on (a) the ability of hyphae to extend from the bead and (b) the degree of sporulation achieved on the bead structure. Knudsen and Bin [83] have considered the effect of incorporating complex nutrients such as bran on the subsequent development of hyphael density of *Trichoderma harzianum* under different soil matrix potentials. They found that hyphal density was significantly higher when bran was incorporated into the bead when exposed to drier soils. Knudsen et al. [85] used the osmotic regulant, polyethylene glycol (PEG), to enhance sporulation of *Beauveria bassiana* and observed an improvement over the conventional pellet formulation. They also found that PEG enhanced hyphal extension of *Trichoderma* spp. in soil. Vestberg and Uosukainen [103] also report enhanced sporulation of *Glomus intraradix* when grown in a hydrogel containing nutrients.

The degree to which cell mobility is restricted within the bead matrix can be influenced by a number of factors. Cells may diffuse passively due to Brownian motion. Certain cultures may be influenced by chemotaxis and motility. However a major mechanism is cell leakage [158, 159]. The rate of leakage appears to be mainly influenced by the growth of the microbial cells – it can be significantly reduced by inhibiting the growth rate of the culture or by altering the gelation procedure to strengthen the matrix [160, 161]. The type of microbial cell will also influence leakage. Outgrowth of single cells result in leakage but outgrowth of mycelia, for example, will cause leakage only if cells are sheared off. Expansion of the matrix pores containing cells has also been observed [162]. Single cells cause gels to decay by the growth and rupture of surface colonies. This removes the gel surface and allows release of cells and nutrient access to successively deeper colonies [163]. Thus the gel decays from the exterior. Tanaka et al. [164] employed a technique involving two layers of alginate, the cells were retained in the inner layer whereas the outer layer prevented leakage. Iijima et al. [165] coated alginate beads with urethane polymer to enhance stability and prevent leakage. Joung and Royer [166] described an expansible cell carrier prepared with a soluble adduct of polyethyleneimine (PEI)-alginate which permitted the expansion of beads to over 1000% of the initial volume as the immobilized cells continued to grow within the carrier gel.

A controlled rate of degradation would mean a controlled release of cells from the protective microenvironment. The biodegradation rate of polymers as a release mechanism is a well developed technology in other areas such as drug release [167]. Bashan [97] has shown that the strength of the alginate beads, the rate of bacterial release and the duration of their survival in soil can be controlled by a variety of gel hardening treatments. When inserted into soil beads with no treatment were completely degraded after 6 weeks whereas those subjected to hardening showed only slightly visible degradation. Both bacteria-free beads and beads containing bacteria with and without skim milk were buried in different soil types – in general it was found that skim milk resulted in higher degradation rates. The presence of bacteria did not appear to affect the breakdown rate significantly.

Thus it appears possible to develop custom designed protection and controlled release systems based on erosion of the polymer surface resulting in the release of physically entrapped cells or through cleavage of covalent bonds between cell walls and polymer followed by diffusion controlled release of the entrapped cells. Furthermore the microenvironments created can enhance the distribution of the target organism through enhanced sporulation or improved extension of hyphae. In environments subject to perturbations the optimal time for successful inoculation is not always well defined, consequently the slow release of the microorganism could ensure a constant inoculation over relatively long periods with an improved ability to exploit environmental opportunities.

The nature of the gel shape (for example, large beads compared with smaller ones) in addition to the gel properties will also influence the rate of cell release and the environment created within the bead.

6 Physical Microenvironments Created on Immobilization in Polymers

The microenvironment created on immobilization within gels is a function of the nature of the gel (which is influenced by the gelation process), the characteristics of the gel shape or bead produced, and the pattern of biomass formation.

6.1 Gel Microenvironment

The majority of studies, to date, have focussed on the physical properties of gels such as mechanical strength, abrasion resistance and diffusion. Less attention has been directed at the microenvironmental or physiological effects arising from the immobilized state.

In general, effects of gel immobilization can be divided into those arising from the properties of the matrix and those related to the biological system either at the level of the whole cell or the individual enzyme systems.

Possible influences include partitioning effects that may occur due to the electrostatic or hydrophilic/hydrophobic interactions with the matrix resulting in microenvironments different from the bulk liquid phase or macrophase. The polymer macromolecules are capable of organizing water and consequently may alter the availability of water, which can result in modifications to metabolism [168]. Growing entrapped colonies may also exert a considerable force through compression on entrapped cells possibly resulting in the dewatering of cells.

Controversy surrounds the properties of water in gels and indeed in microbial cells. Water activity influences inactivation and preservation of microorganisms, it influences enzyme stability, activity and specificity [169, 170]. The effect of the solvent properties of water on survival of microorganisms in polysaccharide gels was investigated by Mugnier and Jung [80] who concluded that an effect was related to some property of water in the biopolymer. The effect of the charged surface of the polyelectrolyte, the porous nature of the gel and the movement of counterions is unclear. Timasheff et al. [171] found that a number of sugars and amino acids were preferentially excluded from solvent adjacent to protein surfaces due to perturbations of solvent structure from contact with the protein structure. Wiggins [172] hypothesised that the zone of low water activity extends for an appreciable distance from the charged surface of biopolymers. These regions can equilibrate with those regions of the pore or interstice within the gel and not adjacent to the polyion (consequently in a state of higher potential) only through altering the molar volume of water because electrostatic forces prevent counter ions from moving away from their polyion. Thus water equilibrates by increasing in density where the concentration of solute is higher and decreasing in density where the concentration of solute is low. The presence of regions of differing activity would obviously have implications for biological systems.

The interaction of physical exclusion due to an expanding biomass linked to mass transfer limitations may result in alteration of the affinity of nutrients by steric hindrance. Other inhibitions may occur through binding of proteins, extracellular polymers or substrates. The influence of ions used in crosslinking polymers must also be considered. Diffusional resistances existing between the different phases (e.g. liquid, solid, gel or gas) may influence mass transfer of substrate or gases from the bulk liquid or in the case of inhibitory end-products, their removal from the metabolizing cells. Alginate gel matrices are negatively charged, consequently pH can influence the diffusion of charged substrates or excretion of products. For example the rate of diffusion of the protein, bovine serum albumin, out of alginate beads increased with increasing pH due to the increased negative charge on the protein [173].

Steenson et al. [174] have concluded that immobilized cultures of *Streptococcus lactis* C2 and *Streptococcus cremoris* HP were protected through exclusion of phage particles from the gel matrix. Champagne et al. [175] reported a similar effect.

The gel matrix can result in a relatively unmixed low shear environment which can influence mixing and mass transfer and provide mechanical support/protection to growing cells.

6.2 Effect of Cells on Gel Microenvironment

The presence of cells may create dynamic conditions, for example, through cell lysis or cell leakage – this may exert a beneficial effect on neighbouring cells. Gel instability resulting from reactions with essential cell nutrients such as phosphate with calcium, which is necessary for alginate stability, can also generate dynamic environmental conditions.

An interaction at a physical level can also occur between microbial cells and matrix – irrespective of viability microbial cells modify and weaken the gel. The presence of cells typically reduces the effective diffusion coefficients in gels [161, 176–179]. The diffusion coefficient is reported to be either proportional to the square of the void fraction in the gel [176, 180] or varies exponentially with the reciprocal of the cell density [124, 181, 182]. Recent work indicates that the cell morphology may also influence diffusivity [183] for example, spiral cells reduced diffusion more than less convoluted straight rod-shaped cells. Gel strength declines with cell loading [122, 140, 161, 184].

6.3. Influence of Bead Diameter on Gel Microenvironment

Key properties influencing the microenvironment within the beads are: (a) the strength and permeability of the gel and (b) bead diameter. An important property of the bead is the surface area to volume ratio and fundamental to the bead microenvironment will be the biomass loading and distribution within the bead. The surface area to volume ratio of the bead is dependent on the bead diameter and will influence diffusion rates and steady-state concentrations.

The following is generally the most popular technique of immobilization. The cells are first suspended in an aqueous solution of gel polymer such as alginate which is then extruded to form drops which are discharged into a curing solution (usually calcium in the case of alginate and potassium in the case of carrageenan). The gel particle diameter is dependent on the capillary used to dispense the drops [185] and the viscosity of the solution which increases with alginate concentration [145]. An alternative approach is described by Nilsson et al. [186] who used a two-phase system – the solution of alginate and cells were added to a hydrophobic phase under stirring, when droplets have been formed the polymer is induced to solidify by cross-linking. The size of the beads could be adjusted between 0.1 and 5 mm in diameter by varying the stirring speed. Although the use of a resonance nozzle [187] or an atomizer results in the production of smaller alginate beads [138, 188], the majority of reports describe beads of 3–5 mm in diameter. This size gives a satisfactory pattern with respect to reactor performance criteria such as fluidisation/settling etc. but results, sometimes, in difficulty in maintaining an evenly distributed biomass throughout the gel matrix.

Dalili and Chau [189] concluded, using mathematical models and reported data, that to maintain a uniform distribution of biomass in beads the bead diameter should be less than 1 mm. This conclusion is reinforced by the numerous reports that cell growth occurs only in the outer 0.3–0.5 mm of the bead. This nonhomogeneous biomass distribution within the immobilization matrix has been reported for immobilized mamalian cells [190], plant [191], algal [122, 192, 193], bacterial cells [163, 194, 195] in addition to yeasts [164, 196, 197, 198] and filamentous fungi [199–202]. From a process aspect there are several significant consequences of this nonhomogeneity in biomass distribution. It leads to zonation of the microenvironment within the bioparticle or bead which may modify intraparticle growth, metabolism, and product formation [203]. Low concentrations of cells at the particle centre will generate significant unproductive regions and may influence particle density. The accumulation of cells at the surface may rupture the gel surface resulting in outgrowth and leakage [123]. Whereas these effects may be undesirable within a controlled fermentation system there may be conditions in which this heterogeneity may be advantageous, for example, if the beads degrade at a defined rate resulting in continuous release of viable cells.

In practice a gel immobilized microbial cell system is heterogeneous in relation to the diffusion of nutrients and products. The distribution and the volume fraction of cells in the bead will affect the diffusivity of solutes. Steady state levels or gradients will exist depending on the balance between the biological demand and the diffusion or supply. Hooijmans et al. [204] using oxygen microsensors investigated oxygen profiles in carrageenan particles containing *E. coli*. The profiles became increasingly steep with time and the oxygen penetration depth within the gel decreased and eventually reached a steady state value of approximately 100 µm. After cell growth occurred the gel became diffusion-limited with the reaction rate exceeding the rate of diffusion. Consequently the oxygen penetration depth is controlled almost totally by the diffusion parameter thus differences in kinetic parameters have only limited effects on the final oxygen penetration depth in microbial systems [205].

However in systems using microbial cultures with relatively low respiration rates the depth would probably be much greater. Plant and hybridoma cells typically grow deeper in the gel, up to 500–1000 µm [206] and 1500–2500 µm [190]. Chen and Humphrey [207] have developed relationships that can be used to estimate the effect of respiration rate on critical particle diameter although their equation is based on the assumption that cell density is uniform throughout.

Monbouqette and Ollis [208] and Burrill et al. [209] observed optimum biomass levels some distance from the surface. Both found the maximum cell concentration about 100 µm below the surface for entrapped *Zymomonas* sp. (in ca-alginate) and *Saccharomyces* sp. (in polyacrylamide) respectively. This effect may be related to the formation of a hard gel layer surrounding the bead.

The obvious answer to the dilemma is to decrease the diameter of the bead to overcome the mass transfer argument although this action will result in a bead with diminished settling properties. For example, the use of immobilization technology in other areas such as biosensor design highlights the fact that the loading capacity of matrices is a balance between the immobilization of sufficient cells to achieve the objective (in this case permit primary signal generation and observation) and the inhibitory effects of excess cells on, for example, oxygen diffusion [210]. However the unique environment consisting of numerous microenvironments along the diameter of the bead – each with cells subjected to different degrees of nutrient limitation, growth rates and presumably differing resistances to stress could have significant applications in the enhanced ecological competence of immobilized inocula.

A number of factors associated with stress resistance and ecological competence have been demonstrated to be influenced by growth rate or energy/nutrient limitation. These include; exocellular polymer levels [45], flocculation [25, 26]; exoenzyme production [211–213]; pigment/antibiotic production [214]; UV resistance [215, 216] and plasmid stability [217].

6.4 Biomass Distribution

The nature of the distribution of a catalyst within the bead can influence the reaction kinetics. Park et al. [218] have shown that biocatalyst performance of the shell configuration is always more effective for immobilized enzymes with positive order reaction kinetics such as Michaelis-Menton and competitive product inhibition whereas in the case of interior "core" configuration negative order reactions such as substrate inhibition is favoured.

Techniques, used in immobilized inocula technology, to alter biomass distribution within alginate beads have included: (a) Biomass immobilized and then stored or used without further culturing. (b) Biomass immobilized then suspended in nutrients and allowed to grow and develop within the matrix before storage and use [97, 102]. (c) Biomass immobilized with complex nutrients (that do not readily diffuse out of the matrix) such as bran, starch, skim milk etc. [97, 100,142].

The difference in these techniques can have a major significance if spores or conidia, for example, are the biomass form immobilized. The potential of the cells, suspended in nutrient or in contact with complex nutrients, to recover from stress arising from the immobilization procedure must also be greater.

A number of process factors will influence biomass distribution. The concentration of polymer within the gel will limit the diffusion rate of solutes. Diffusion will reflect the pore structure, higher concentrations of polymers will hinder diffusion particularly for high molecular weight compounds. This has been demonstrated in alginate gels for glucose and ethanol diffusion [179, 219]. The nature of the matrix polymer (hydrophobicity, solvophobicity etc.) will also regulate diffusion and ultimately biomass distribution.

Furthermore the distribution of biomass and the steady state levels of nutrient solutes will be altered by cell loading. It has been reported that the inoculation of gels with low levels of biomass and subsequent growth within the bead improves the quality of the biocatalyst [158]. The effect of inoculum loading on productivity of immobilized systems has received considerable attention and a number of workers have commented on its effect on growth. There is less information on its effect on biomass distribution. Mussenden et al. [220] have described how *Penicillium chrysogenum* spore loading in k-carrageenan influenced the biomass loading and the size of the subsequent bioparticle (Fig. 10). Leo [142] has demonstrated that hyphal biomass produced per viable centre was higher at low inoculum loadings. Many workers have found no difference: the inoculum size did not influence the dry weight of *Aspergillus niger* [221] *A. terreus* [199] or *S. cerevisiae* [197]. However early differences in growth with inoculum loading are probably not sustained in older bioparticles if they approach the same biomass content.

Some comparative studies of biomass distribution with different strains of immobilized cells indicated that morphology and distribution can be strain dependent. El-Sayed and Rehm [202] made a qualitative comparison of two strains of *P. chrysogenum* – one strain resulted in more pronounced and persistent microcolonies. Leo [142] has examined the distribution of *P. chrysogenum* biomass as reflected by protein levels within the alginate bead suspended in relatively rich nutrient environments such as medium. He used the 75% depth as an index – this is the depth above which 75% of the biomass occurs. He has shown that it varies with strain and that it tends to move markedly towards the

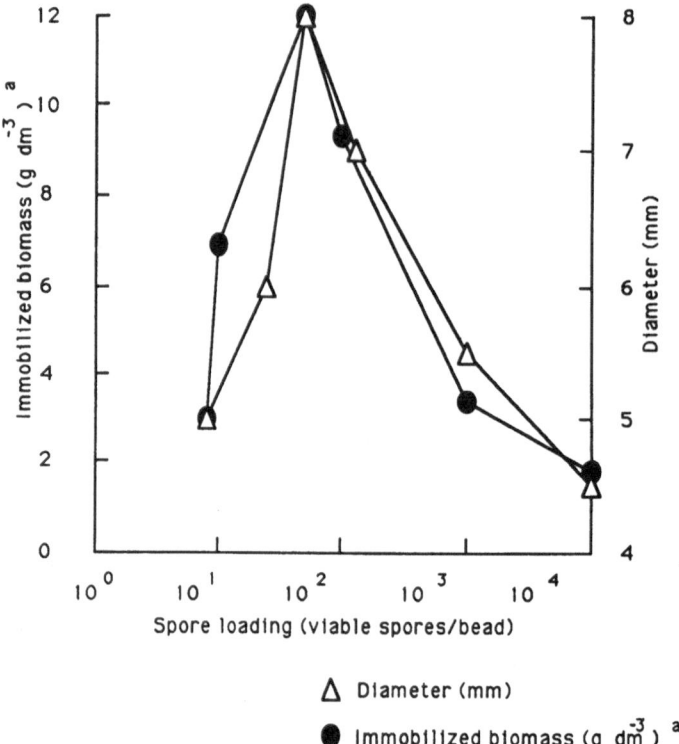

Fig. 10. The influence of spore loading on biomass patterns and bead characteristics. (data from Ref. 220). Initial bead diameter = 3.7–4.0 mm.
[a]Concentration in g dm^{-3} of reactor working volume

surface as a consequence of outgrowth of hyphae from the particle. This is a fundamental difference between single cells and mycelial types which also have the ability to translocate nutrients.

Berry et al. [222] described large microcolony development when small levels of inoculum were used whereas growth was restricted to the outer layer of the bead when large numbers of bacteria were used as inoculum.

Distribution of biomass can be manipulated by growing the bioparticle in nutrient limited conditions. More uniform distribution and reduced outgrowth have been achieved under nitrogen and phosphate limitation. Growth of *A. niger* was peripheral using 0.1–0.2 g l^{-1} NH_4NO_3 but more uniform and sparse with 0.05 g l^{-1} [223]. Similarly for *A. niger* in calcium alginate phosphate levels of 0.075–0.01 g KH_2PO_4 per l caused a shift to peripheral growth which was more pronounced at 0.125–0.15 g l^{-1} [200]. The reason for the improved distribution appears to be related to the lower overall biomass loading in the particle after a period of incubation nutrient access to the core of the bead is greater, consequently central biomass formation is higher.

Examples of nutrient entrapment have been reported. Leo [142] found improved biomass distribution when complex nutrients were incorporated into the bead. O'Riordan et al. [224] used co-immobilization of chitin to regulate chitinase activity in calcium alginate beads. Algal cells have been entrapped in alginate which was dissolved in growth medium [203]. Another approach was that of Mak and Trevan [225] who described a system in which urease was co-immobilized with *Chlorella* sp. to improve biomass retention within the bead by supplying an internal source of carbon and nitrogen. This increased the depth of cell growth based on the depth of gas evolution as seen on electron micrographs.

Other studies demonstrated the importance of mass transfer in developing biomass distributions. Uniformity of biomass distribution was increased in the initial stages of development when *Penicillium urticae* was entrapped in carrageenan containing open pores formed by incorporation of air at gelation [226]. Sutar et al. [227] produced similar type open pore agar particles by incorporating alginate into agar droplets and dissolving out the alginate after gellation.

Other environmental parameters interacting with biomass distribution and activity would include pH. Although the importance of pH in the micro-environment is obvious it is difficult to make reliable pH measurements within the bead due to the unacceptable level of disturbance which this entails. Gradients have been observed in alginate beads during the cleavage of Penicillin G by immobilized *E. coli* [228]. Control of the gradients is also difficult. One strategy involves the incorporation of tricalcium phosphate – its solubility at neutral pH is low, its dissolution at acid pH values prevent the development of acid gradients [229]. The use of insoluble $CaCO_3$ both as a source of calcium ions and as a proton acceptor coupled to the use of glucono-delta-lactone (GDL) was another strategy used to regulate intra-bead pH [230].

Liou and Rousseau [231] describe another approach – a coupled system involving an enzymatic reaction to balance any pH shift due to the main reaction. The use of protein sources such as skim milk [97, 99] would possibly contribute to significant pH regulation within beads in which acid was produced.

A major technique used to modify the bead microenvironment is co-immobilization. Co-immobilization of algae with bacteria has been used to overcome oxygen diffusional resistances and to improve oxygen supply to the bacterial population, [232–234]. Co-immobilization allows nitrification and denitrification to occur simultaneously within the same system due to the presence of aerobic and anaerobic regions in the matrix [127, 235]. Defined consortia of immobilized cells have been used to direct substrate breakdown and/or product formation [28, 236, 237].

The design of inocula consisting of consortia of interacting microorganisms obviously offers great potential but will require a much greater understanding of ecological principles before it can be applied in inoculum technology. However it does probably represent an advanced application of the principle of microenvironment development. Undesirable consortia or alterations to the bead microenvironment caused by contaminating microorganisms can be minimized by the incorporation of selective bacteriostatic agents [238].

Thus it is evident that the microenvironment created on immobilization can be successfully regulated. Furthermore it can be manipulated to allow controlled release of the immobilized cells. However the exploitation of the unique properties of such an environment is yet almost unexplored. Furthermore immobilization has major influences on microbial metabolism. The following section will deal with some of the evidence for the ecophysiological influences exerted by immobilization.

7 Effects of Gel Immobilization on Cell Physiology

In the past a major limitation in studies on the physiology of immobilized cells arose from the absence of suitable in situ noninvasive techniques. Recent studies based on spectroscopic techniques and microfluorimetry have provided direct evidence that immobilization does in fact alter metabolism which is consistent with earlier reports on differences between immobilized and suspended cell behaviour.

There is little doubt that cell immobilization in gels can exert significant qualitative changes in cellular function and metabolism which consequently indicates that the technique has significant potential as a means of controlling and manipulating physiology. For example, immobilization has been observed to fundamentally alter metabolism in a variety of cells which include changes in morphology [203, 239, 240] yield of product [210, 241, 242] and kinetics of growth and product formation [243, 244]. Polymers such as alginate have been shown to enhance enzyme activity in, for example, *Pseudomonas* [245, 246]. Sodium alginate has been shown to alter growth patterns, for example, inhibit pellet formation in submerged cultures of *Rhizopus arrhizus* [247].

A number of reports have provided evidence that immobilized microbial cells do behave differently compared with free cells. Galazzo and Bailey [248–250] examined the metabolic control exerted through immobilization. Yeast cells immobilized in calcium alginate had substantially altered relative fluxes in the different transport and enzyme pathways involved in the metabolism of glucose to ethanol. For example, the analysis of glucose, ethanol and glycerol formation data showed that the immobilized cells possessed phosphofructokinase, ATPase and polysaccharide synthesis activities which were approximately two-, two-, and nine-fold greater respectively than observed in free cells. The largest flux control for suspended cells occurred at the glucose uptake step while the greatest flux control for immobilized cells was exerted at the reaction step catalysed by the enzyme phosphofructokinase. Furthermore the reactions directed towards polysaccharide synthesis assumed significant flux control in immobilized cells whereas they were relatively insignificant in suspended cells. The enhanced synthesis of glycogen observed in immobilized cells would lead to improved ecological competence arising from such storage products.

Ryu et al. [251] attempted to estimate the basic intrinsic process kinetics of immobilized cells. They examined substrate saturation constants, maximum reaction rates, substrate inhibition constants and maximum ethanol tolerance level. They found that these constants varied with immobilization conditions such as gel particle diameter. Monbouquette et al. [252] have shown that RNA levels per cell in cells near the bead surface were higher relative to those deeper within the gel bead confirming a radial gradient in immobilized cell physiology/activity.

Hilge-Rotmann and Rehm [253] compared physiological properties of alginate-entrapped cells of *S. cerevisiae* with free cells under similar cultural conditions. Cells grew in microcolonies in the alginate beads and showed improved glucose uptake and ethanol productivity with simultaneous decreased cell yields. Cells in microcolonies showed an increased specific hexokinase and phosphofructokinase activity whereas single immobilized cells showed only very slight effects which suggested that the alterations in physiology are connected with growth of cells in microcolonies or aggregates.

The level of extracellular polymer associated with cells appears to be important in developing ecological competence and appears to be influenced by immobilization. El-Sayed et al. [254] found that immobilized cells of *Leuconostoc mesenteroides* produced more dextran than free cells under otherwise comparable conditions. Wang and Wang [255] investigated the mechanism for biopolymer accumulation in immobilized *Acinetobacter calcoaceticus* systems and found that cells subjected to zero shear stress (such as present in the carrier matrix) could produce cell bound polymer to dry weight cell ratio as high as 1.6. However at higher shear levels (for example, above 0.5 Pa – a level equivalent to the shear experienced by free cells in a stirred tank reactor) cell–bound polymer to dry weight ratio decreased to 0.2.

Cell division appears to be influenced by immobilization. Jirkü et al. [239] showed from electron microscopic studies that cell division proceeded without separation of the daughter cells resulting in chain like filaments indicating a degree of polarization of budding. Doran and Bailey [243] observed high numbers of polyploid cells when *S. cerevisiae* was immobilized which suggested a disruption of the normal cell cycle, and the possibility exists that polyploid daughter cells arose as a result of intracellular crowding due to immobilization.

Immobilization also confers protection to cells exposed to toxic or inhibitory substrates/environments and alginate beads have been reported as providing microenvironments which enhance the degradation of xenobiotics. Beunink and Rehm [256] describe the use of co-immobilized *Alcaligenes* sp. and *Enterobacter cloacae* in alginate beads to allow the simultaneous reductive dechlorination of

1,1,1-trichloro-2,2-bis(4-chlorophenyl)ethane (DDT) and the oxidative degradation of 4,4'-dichlorodiphenylmethane (DDM), thus showing that complex
aerobic and anaerobic processes can be carried out simultaneously by immobilized cultures and indeed exploits one of the mass transfer limitations associated
with the use of immobilized systems employing pure cultures. Kutney et al.
[257] have reported the use of alginate beads in the transformation of dehydro-
abietic acid (DHA) into nontoxic metabolites by the zygomycete *Mortierella
isabellina*. They found that immobilization resulted in greater long term stability
of enzyme activity compared with free mycelia and that the breakdown products
differed which they attributed to a greater level of secondary metabolism.

Bettman and Rehm [258] reported that immobilized cells of *Pseudomonas
putida* entrapped in calcium alginate could be exposed to higher phenol
concentrations without loss of viability. Keweloh et al. [259] found that immobilization of cells in calcium alginate beads reduced the growth inhibition caused
by bacteriostatic concentrations of phenol but the degree of protection was
found to correlate with the formation of microcolonies in the gel matrix.
Keweloh et al. [260] reported that immobilization of *E. coli* cells altered the
characteristics of the membrane and resulted in marked changes in the protein
pattern of the outer membrane.

Immobilization of cells offered advantage when membrane interacting compounds induced a loss of cellular ions and metabolites. Heipieper et al. [261]
observed that cells immobilized and grown in alginate suffered a smaller loss of
cations when exposed to phenols. The reestablishment of gradients was observed at a higher concentration of the phenol with immobilized cells compared
with free cells and less membrane damage was observed, thus leading to
increased phenol tolerance. These workers hypothesized that after the loss of
metabolites due to an increase in permeability the cells are faced with relatively
high concentrations of these compounds inside the microcolonies. Thus in
contrast to free cells, the instantaneous dilution or loss of material into the gel-
free medium is prevented and reuptake can be achieved easily.

Evidence from studies on protoplast behaviour demonstrate beneficial
effects of gels on cell recovery after stress. It has been shown that yeast
protoplasts immobilized in alginate showed improved ability to regenerate and
revert to cells [262]. A possible explanation is that the gel prevents the
diffusion of the more soluble fractions of the cell wall such as glycoproteins. In
liquid media the surface of the protoplast retains only the glucan fibrillar
components and releases the amorphous glycoproteins into the media [263].
Schnabl et al. [264] observed that entrapped protoplasts of *Vicia faba* were
more stable to mechanical and osmotic stress than freely suspended protoplasts
which has led to the suggestion that the technique may have application in the
transport and storage of protoplasts. Linsefors and Brodelius [265] examined a
number of gels formed from natural polymers and showed that immobilized
protoplasts of plant cells showed a much higher viability after 14 days as
compared with free protoplasts under the same conditions. They also showed

that immobilized protoplasts can withstand various stress arising from process operations (e.g. washing and centrifugation) better than free ones.

Thus, in general, there is abundant evidence that immobilization can alter microbial metabolism and response resulting in enhanced stability and can also protect the immobilized cells from adverse influences.

Evidence from a number of studies on a variety of immobilized cell systems indicate that the effects of immobilization can be maintained for some time after cell release. Mori et al. [266] observed that newly released cells of *Acetobacter* sp. from k-carrageenan had an improved level of activity and retained it for a few generations after leakage. On release the specific growth rate was 1.6 times higher than that of non-immobilized cells and the respiratory activity was three times greater. Studies on plant cell immobilization have shown that cells when released grow in fine cell suspensions for a number of generations with the absence of cell aggregation which occurs with free cell suspensions [267–269] which indicates that the effect of immobilization is a fundamental effect and is retained for a number of generations after release.

8 Conclusions

The largest and most widespread examples of biotechnological process intensification and stabilization by immobilization are in the case of aerobic wastewater treatment, activated sludge and trickling filtration processes and in the case of anaerobic wastewater treatment, the upflow anaerobic sludge blanket process [270]. All of the above are effectively open, continuous, mixed populations in which immobilization processes, like flocculation or biofilm growth, are employed to overcome fluctuations arising from the mixed nature of the process culture, variations in the environmental conditions encountered, and the diversity of substrates and nutrients available. However the verifiable information concerning the ecophysiology of immobilized cells is still scant. Other heterogeneous environments, especially those requiring the addition of inocula have received less attention from the bioengineer. However a combination of approaches based on microbial ecology and process engineering will be necessary for further exploitation of such environments.

The combination of nutrient limitation and immobilization technology has resulted in the development of inocula delivery systems which improve the resistance of the culture and regulate the release of cells from the protective microenvironment. The enhanced ecological competence arising from the higher level of stress resistance can increase the ability to compete within the macroenvironment and the controlled release can ensure a continuous inoculation over time which has obvious advantageous in dynamic environments where the "window of opportunity" to allow successful colonization can be very restricted.

The further incorporation of both space and time dimensions into biological processing offers exciting opportunities. The creation of microenvironments in both space and time dimensions will obviously contribute to the further regulation of stability/instability of microbial systems. Thus, for example, the technology offers the potential to control the instability or alternatively the persistence of genetically engineered microbes when released into natural environments.

The use of inexpensive, naturally occurring polymers offers unique advantages to create protective microniches based on the gels ability to bind water, partition ions and molecules, exclude predators and provide a relatively unmixed region surrounding the cell. The juxtapositioning created through co-immobilization of cells, cells plus protectants or selective agents and/or cells plus nutrients offers great potential in the creation and development of microniches.

The characteristics of the gel and its shape has a major influence on the microniche created. Through the gelation process the rate of diffusion of molecules or the rate of polymer breakdown (or cell release) can be regulated. Surface area to volume ratio, or in the case of bead shapes the diameter, can influence the biomass distribution throughout the gel. The radial gradients created (and the resulting degree of nutrient limitation) in gel beads can have a significant influence on the behaviour of the immobilized biomass.

The combination of the influences exerted through immobilization has significant effects on microbial metabolic control. Studies have demonstrated a significant effect on flux control with enhanced synthesis of reserve polymers and enhanced exopolymer production and altered membrane function. The development of microcolonies within the bead structure would appear to exert beneficial effects in protecting cells against stress.

9 References

1. Tempest DW, Neijssel OM, Zevenboom W (1983) In: Slater JH, Whittenbury R, Wimpenny JWT (eds) Microbes in their natural environments. Cambridge University Press, Cambridge, p 119
2. Smith RL, Schank SC, Milam JR, Baltensperger AA (1984) Appl Environ Microbiol 47:1331
3. Okon Y, Hadar Y (1987) CRC Crit Rev Biotechnol 6:61
4. Jutsum AR (1988) Phil Trans R Soc Lond B 318:357
5. Campbell R, Macdonald RM (1989) Microbial inoculation of crop plants. IRL Press, Oxford
6. Nesbakken T, Broch-Due M (1991) J Sci Food Agric 54:177
7. Berg HC (1975) Nature 254:389
8. Lauffenburger DA, Rivero M, Kelly F, Ford R, DiRienzo J (1987) Ann NY Acad Sci 506:281
9. Purcell EM (1977) Am J Phys 45:3
10. Berg HC, Purcell EM (1977) Biophys J 20:193
11. Kelly FX, Dapsis KJ, Lauffenburger DA (1988) Microb Ecology 14:115
12. Brown DA, Berg HC (1974) Proc Natn Acad Sci USA 71:1388
13. Adler J (1969) Science 166:1588
14. Adler J (1973) J Gen Microbiol 74:77

15. Adler J, Templeton B (1967) J Gen Microbil 46:175
16. Adler J, Hazelbauer GL, Dahl MM (1973) J Bacteriol 115:824
17. Gannon JT, Manilal VB, Alexander M (1991) Appl Environ Microbiol 57:190
18. Allee WC, Emerson AE, Park O, Park T, Schmidt KP (1949) Principles of animal ecology. Saunders WB, Philadelphia
19. Karel SF, Libicki SB, Robertson CR (1985) Chem Eng Sci 40:1321
20. Brohan B, McLoughlin AJ (1984) Appl Microbiol Biotechnol 20:10
21. Brohan B, McLoughlin AJ (1984) Appl Microbiol Biotechnol 20:16
22. Brohan B, McLoughlin AJ (1984) Appl Microbiol Biotechnol 20:146
23. Li DH, Ganczarczyk JJ (1986) CRC Crit Rev Envir Control 17:53
24. McKinney RE (1952) Sewage Ind Wastes 24:280
25. Brohan B, McLoughlin AJ (1985) Ir J Fd Sci Technol 9:43
26. McLoughlin AJ, Vallom JK (1984) J Appl Bacteriol 57:485
27. Pavoni JL, Tenney MW, Echelberger WF (1972) J Wat Pollut Control Fed 44:414
28. Jones WJ, Guyot J-P, Wolfe RS (1984) Appl Environ Microbiol 47:1
29. Conrad R, Phelps TJ, Zeikus JG (1985) Appl Environ Microbiol 50:595
30. Thiele JH, Chartrain M, Zeikus JG (1988) Appl Environ Microbiol 54:10
31. Thiele JH, Zeikus JG (1988) Appl Environ Microbiol 54:20
32. Crombie-Quilty MB, McLoughlin AJ (1983) Water Res 17:39
33. McLoughlin AJ, Crombie-Quilty MB (1983) Biotechnol Bioeng 25:2905
34. McLoughlin AJ, Crombie-Quilty MB (1983) Water Res 17:161
35. Doyle RJ, Chaloupka J, Vinter V (1988) Microbiol Rev 52:554
36. Goodell EW, Higgins CF (1987) J Bacteriol 169:3861
37. Greenway DLA, Perkins HR (1985) J Gen Microbiol 131:253
38. Azam F, Ammerman JW (1984) In: Fasham MJR (ed) Flows of energy and material in marine ecosystems. Plenum, New York, p 345
39. Jackson GA (1987) In: Fletcher M, Gray TRG, Jones JG (eds) Ecology of microbial communities. Cambridge University Press, Cambridge, p 213
40. Mason CA, Bryers JD, Hamer G (1986) Chem Eng Commun 45:163
41. Mason CA, Hamer G, Bryers JD (1986) FEMS Microbiol Rev 39:373
42. Tenney MW, Stumm W (1965) J Wat Pollut Control Fed 39:1370
43. Busch PL, Stumm W (1968) Envir Sci Technol 2:49
44. Nishikawa S, Kuriyama M (1968) Water Res 2:811
45. Valom K, McLoughlin AJ (1984) Water Res 18:1523
46. Allison FE (1968) Soil Sci. 106:136
47. Wilkinson JF (1958) Bacteriol Rev 22:46
48. Harris RH, Mitchell R (1973) Ann Rev Microbiol 27:27
49. Dudman WF (1977) In: Sutherland I (ed) Surface carbohydrates of the prokaryotic cell. Academic, London, p 357
50. Martens DA, Frankenberger WT (1991) Soil Biol Biochem 23:731
51. Atkinson B, Daoud IS (1976) Adv Biochem Eng 4:41
52. Atkinson B, Fowler HW (1974) Adv Biochem Eng 3:221
53. Danso SKA, Keya SO, Alexander M (1975) Can J Microbiol 21:884
54. Habte M, Alexander M (1977) Arch Microbiol 113:181
55. Clarholm M (1981) Microb Ecology 7:343
56. Senoo K, Nishiyama M, Wada H, Matsumoto S (1992) FEMS Microb Ecology 86:311
57. Lewis JA, Papavizas GC (1985) Plant Pathol 34:571
58. Burton JC (1979) In: Peppler HJ, Perlman D (eds) Microbial Technology, Academic, New York, vol 1, p 29
59. Graham-Weiss L, Bennett ML, Paau AS (1987) Appl Environ Microbiol 53:2138
60. Chao W-L, Alexander M (1984) Appl Environ Microbiol 47:94
61. Philpotts H (1976) J Appl Bacteriol 41:277
62. Grunewaldt-Stocker G (1989) Agric Ecosystems Environ 29:179
63. Kandasamy R, Prasad NN (1971) Curr Sci 40:496
64. Jones RW, Pettit RE, Taber RA (1984) Phytopathology 74:1167
65. Saito M (1989) Agric Ecosystems Environ 29:341
66. Hamdi YA, Al-Tai AM, Khazal R, Abbas HI (1982) Egypt J Microbiol 17:15
67. Flemming CA, Ferris FG, Beveridge TJ, Bailey GW (1990) Appl Environ Microbiol 56:3191
68. Postma J, Hok-A-Hin CH, van Veen JA (1990) Appl Environ Microbiol 56:495
69. Postma J, Walter S, van Veen JA (1989) Soil Biol Biochem 21:437

70. Heijnen CE, van Veen JA (1991) FEMS Microbiology Ecology 85:73
71. Hattori T, Hattori R (1976) CRC Crit Rev Microbiol 4:423
72. Darbyshire JF (1976) J Soil Sci 27:369
73. Bashan Y (1986) Appl Environ Microbiol 51:1067
74. Opara CC, Mann J (1988) Biotechnol Bioeng 31:470
75. Messing RA, Oppermann RA (1979) Biotechnol Bioeng 21:49
76. Messing RA, Oppermann RA, Kolot FB (1979) Biotechnol Bioeng 21:59
77. Klein J, Ziehr H (1990) J Biotechnol 16:1
78. Dommergues YR, Diem HG, Divies C (1979) Appl Environ Microbiol 37:779
79. Jung G, Mugnier J, Diem HG, Dommergues YR (1982) Plant Soil 65:219
80. Mugnier J, Jung G (1985) Appl Environ Microbiol 50:108
81. Walker HL, Connick WJ (1983) Weed Sci 31:333
82. Fravel DR, Marois JJ, Lumsden RD, Connick WJ (1985) Phytopathology 75:774
83. Knudsen GR, Bin L (1990) Phytopathology 80:724
84. Knudsen GR, Eschen DJ (1991) Plant Disease 75:466
85. Knudsen GR, Eschen DJ, Dandurand LM, Wang ZG (1991) Appl Environ Microbiol 57:2864
86. Lewis JA, Papavisas GC (1987) Plant Pathol. 36:438
87. Lewis JA, Papavisas GC (1988) Phytopathology 78:862
88. Lumsden RD, Lewis JA (1989) In: Whipps JM, Lumsden RD (eds) Biotechnology of fungi for improving plant growth. Cambridge University Press, Cambridge, p 171
89. Lumsden RD, Locke JC (1989) Phytopathology 79:361
90. Papavizas GC, Fravel DR, Lewis JA (1987) Phytopathology 77:131
91. Mauperin Ch, Mortier F, Garbaye J, Le Tacon F, Carr G (1987) Can J Bot 65:2326
92. Le Tacon F, Jung G, Mugnier J, Michelot P, Mauperin C (1985) Can J Bot 63:1664
93. Passmore N (1989) Australian Hort. Feb. 68
94. Kropáček K, Cudlín P, Mejstřík V (1989) Agric Ecosystems Environ 28:263
95. Strullu D-G, Plenchette C (1990) CR Acad Sci Paris 310:447
96. Sylvia DM, Jarstfer AG (1992) Appl Environ Microbiol 58:229
97. Bashan Y (1986) Appl Environ Microbiol 51:1089
98. Fages J (1990) Appl Microbiol Biotechnol 32:473
99. Kearney L, Upton M, McLoughlin AJ (1987) Ir J Fd Sc Technol 11:182
100. Kearney L, Upton M, McLoughlin AJ (1990) Appl Environ Microbiol 56:3112
101. Kearney L, Upton M, McLoughlin AJ (1990) Appl Microbiol Biotechnol 33:648
102. Kuek C, Tommerup IC, Malajczuk N (1992) Mycol Res 96:273
103. Vestberg M, Uosukainen M (1992) The Mycologist 6:38
104. Kumar PKR, Schügerl K (1990) J Biotechnol 14:255
105. Küster E, Rodgers J, McLoughlin AJ (1968) In: Lafleur C, Butler J (eds) Proc 3rd International Peat Congress, Runge Press, Ottawa, p 23
106. McLoughlin AJ, Küster E (1972) Proc R Ir Acad 72B:1
107. McLoughlin AJ, Küster E (1972) Plant and Soil 37:17
108. Mitchell AF (1972) US Patent No 3 649 239
109. Scher HB (1977) US Patent No 4 053 627
110. Connick WJ (1983) US Patent No 4 401 456
111. Fages J, Mulard D (1986) French Patent No 2 600 673
112. Lewis JA, Papavizas GC (1987) US Patent No 4 668 512
113. Marois JJ, Fravel DR, Connick WJ, Walker HL, Quimby PC (1988) US Patent No 4 724 147
114. Woodward J (1988) J Microbiol Methods 8:91
115. Gilson CD, Thomas A, Hawkes FR (1990) Process Biochemistry International 25:104
116. Grant GT, Morris ER, Rees DA, Smith PJC, Thom D (1973) FEBS Letters 32:195
117. Veelken M, Pape H (1984) Appl Microbiol Biotechnol 19:146
118. Ak MM, Nussinovitch A, Campanella OH, Peleg M (1989) Biotech Progress 5:75
119. Rochefort WE, Rehg T, Chau PC (1986) Biotechnol Letters 8:115
120. Vidoli R, Yamazaki H, Nasim A, Veliky IA (1982) Biotechnol Letters 4:781
121. Paul F, Vignais PM (1980) Enzyme Microb Technol 2:281
122. Dainty AL, Goulding KH, Robinson PK, Simpkins I, Trevan MD (1986) Biotechnol Bioeng 28:210
123. Tanaka H, Irie S (1988) Biotechnol Techniques 2:115
124. Klein J, Manecke G (1982) In: Chibata I, Fukui S, Wingard LB (eds) Enzyme Engineering. Plenum, New York, vol 6, p 181
125. Gray CJ, Dowsett J (1988) Biotechnol Bioeng 31:607

126. King AH (1983) In: M Glicksman (ed) Food Hydrocolloids. CRC Press Inc., Florida, vol 2, p 115
127. Kokufuta E, Yukishige M, Nakamura I (1987) J Ferm Technol 65:659
128. Leo WJ, McLoughlin AJ, Malone DM (1990) Biotechnol Prog 6:51
129. Robinson PK, Goulding KH, Mak AL, Trevan MD (1986) Enzyme Microb Technol 8:729
130. Tanaka H, Matsumura M, Veliky IA (1984) Biotechnol Bioeng 26:53
131. Knorr D, Daly M (1988) Process Biochem 23:48
132. Kierstan M, Bucke C (1977) Biotechnol Bioeng 19:387
133. Fang BS, Fang HY, Wu CS, Pan CT (1983) Biotechnol Bioeng Symposium 13:457
134. Ketel DH, Hulst AC, Gruppen H, Breteler H, Tramper J (1987) Enzyme Microb Technol 9:303
135. Musgrave SC, Kerby NW, Codd GA, Stewart WDP (1983) Eur J Appl Microbiol Biotechnol 17:133
136. Scherer P, Kluge M, Klein J, Sahm H (1981) Biotechnol Bioeng 23:1057
137. Kuek C, Armitage TM (1985) Enzyme Microb Technol 7:121
138. Klein J, Stock J, Vorlop K-D (1983) Eur J Appl Microbiol Biotechnol 18:86
139. Casson D, Emery AN (1987) Enzyme Microb Technol 9:102
140. Krouwel PG, Harder A, Kossen NWF (1982) Biotechnol Letters 4:103
141. Skjåk-Braek G, Grasdalen H, Draget KI, Smidsrød O (1990) In: Crescenzi V, Dea ICM, Paoletti S, Stivala SS, Sutherland I (eds) Recent developments in industrial polysaccharides. Gordon and Breach, New York, p 345
142. Leo WJ (1990) The influence of techniques of immobilization on growth and secondary metabolite production by mycelial organisms. Ph.D. Thesis, National University of Ireland
143. Martinsen A, Skjåk-Braek G, Smidsrød O (1989) Biotechnol Bioeng 33:79
144. Birbaum S, Pendleton R, Larsson P-O, Mosbach K (1981) Biotechnol Letters 3:393
145. Johansen A, Flink JM (1986) Enzyme Microb Technol 8:737
146. Shirai Y, Hashimoto K, Irie S (1989) Appl Microbiol Biotechnol 31:342
147. Guiseley KB (1989) Enzyme Microb Technol 11:706
148. Wang HY, Lee SS, Takach Y, Cawthon L (1982) Biotechnol Bioeng Symp No 12:139
149. Gacesa P (1988) Carbohydr Polym 8:161
150. Wainwright M, Sherbrock-Cox V (1981) Bot Mar 24:489
151. Hansen JB, Doubet RS, Ram J (1984) Appl Environ Microbiol 47:704
152. Gacesa P, Wusteman FS (1990) Appl Environ Microbiol 56:2265
153. Linker A, Evans LR (1984) J Bacteriol 159:958
154. Kondrat'eva LM, Mun TK, Vakhrusheva EV (1988) Mikrobiologiya 57:47
155. Sutherland IW, Keen GA (1981) J Appl Biochem 3:48
156. Boyd J, Turvey JR (1978) Carbohydr Res 66:187
157. Brown BJ, Preston JF, Ingram LO (1991) Appl Environ Microbiol 57:1870
158. Wada M, Kato J, Chibata I (1980) Eur J Appl Microbiol Biotechnol 10:275
159. King VA-E, Zall RR (1983) Process Biochemistry 18:17
160. Nilsson I, Ohlson S (1982) Eur J Appl Microbiol Biotechnol 14:86
161. Cheetham PSJ, Blunt KW, Bucke C (1979) Biotechnol Bioeng 21:2155
162. Siess MH, Divies C (1981) Eur J Appl Microbiol Biotechnol 12:10
163. Nasri M, Sayadi S, Barbotin J-N, Dhulster P, Thomas D (1987) Appl Environ Microbiol 53:740
164. Tanaka H, Irie S, Ochi H (1989) J Ferm Bioeng 68:216
165. Iijima S, Mano T, Taniguchi M, Kobayashi T (1988) Appl Microbiol Biotechnol 28:572
166. Joung JJ, Royer GP (1990) Ann NY Acad Sci 589:271
167. Pitt CG, Schindler A (1983) In: Bruck SD (ed) Controlled Drug Delivery. CRC Press Inc., Florida. vol 1, p 53
168. Mattiasson B, Hahn-Hägerdahl B (1982) Eur J Appl Microbiol Biotechnol 16:52
169. Hahn-Hägerdal B (1986) Enzyme Microb Technol 8:322
170. Hahn-Hägerdal B (1990) CRC Crit Rev Biotechnol 9:259
171. Timasheff SN, Arakawa T, Inoue H, Gekko K, Gorbunoff MJ, Lee JC, Na GC, Piltz EP, Prakash V (1982) In: Franks F, Mathias S (eds) Biophysics of water. J Wiley, Chichester. p 48
172. Wiggins PM (1990) Microbiol Rev 54:432
173. Martinsen A, Storro I, Skjåk-Braek G (1992) Biotechnol Bioeng 39:186
174. Steenson LR, Klaenhammer TR, Swaisgood HE (1987) J Dairy Sci 70:1121
175. Champagne CP, Girard F, Morin N (1988) Biotechnol Letters 10:463
176. Furusaki S, Seki M (1985) J Chem Eng Japan 18:389
177. Furui M, Yamashita K (1985) J Ferm Technol 63:167

178. Sakaki K, Nozawa T, Furusaki S (1988) Biotechnol Bioeng 31:603
179. Scott CD, Woodward CA, Thompson JE (1989) Enzyme Microb Technol 11:258
180. Sun Y, Furusaki S, Yamauchi A, Ichimura K (1989) Biotechnol Bioeng 34:55
181. Larroche C, Gros J-B (1989) Biotechnol Bioeng 34:30
182. Mignot L, Junter G-A (1990) Appl Microbiol Biotechnol 32:418
183. Mignot L, Junter G-A (1990) Appl Microbiol Biotechnol 33:167
184. Cheetham PSJ (1979) Enzyme Microb Technol 1:183
185. Fildes P, Smart WAM (1926) British J Exp Pathology 7:68
186. Nilsson K, Birnbaum S, Flygare S, Linse L, Schröder U, Jeppsson U, Larsson P-O, Mosbach K, Brodelius P (1983) Eur J Appl Microbiol Biotechnol 17:319
187. Hulst AC, Tramper J, van't Riet K, Westerbeek JMM (1985) Biotechnol Bioeng 27:870
188. Rehg T, Dorger C, Chau PC (1986) Biotechnol Letters 8:111
189. Dalili M, Chau PC (1987) Appl Microbiol Biotechnol 26:500
190. Al-Rubeai M, Spier R (1989) Appl Microbiol Biotechnol 31:430
191. Tanaka A Sonomoto K, Usui N, Nakajima H, Fukui S, Fumihiko S, Yamada Y (1984) In: Proc 3rd European Congress on Biotechnology. Verlag-Chemie, Weinheim, vol 1, p 287
192. Robinson PK, Dainty AL, Goulding KH, Simpkins I, Trevan MD (1985) Enzyme Microb Technol 7:212
193. Trevan MD, Mak AL (1988) Tibtech 6:68
194. Shinmyo A, Kimura H, Okada H (1982) Eur J Appl Microbiol Biotechnol 14:7
195. De Taxis due Poët P, Arcand Y, Bernier R, Barbotin J-N, Thomas D (1987) Appl Environ Microbiol 53:1548
196. Anselme MJ, Tedder DW (1987) Biotech Bioeng 30:736
197. Godia F, Casas C, Castellano B, Sola C (1987) Appl Microbiol Biotechnol 26:342
198. Chen KC, Huang C-T (1988) Enzyme Microb Technol 10:284
199. Kautola H, Vahvaselka M, Linko Y-Y, Linko P (1985) Biotechnol Letters 7:167
200. Honecker S, Bisping B, Yang Z, Rehm H-J (1989) Appl Microbiol 31:17
201. Kopp B, Rehm H-J (1984) Appl Microbiol Biotechnol 19:141
202. El-Sayed A-HMM, Rehm HJ (1986) Appl Microbiol Biotechnol 24:89
203. Shi D-J, Brouers M, Hall DO, Robins RJ (1987) Planta 172:298
204. Hooijmans CM, Briasco CA, Huang J, Geraats BGM, Barbotin J-N, Thomas D, Luyben KChAM (1990) Appl Microbiol Biotechnol 33:611
205. Huang J, Hooijmans CM, Briasco CA, Geraats SGM, Luyben KChAM, Thomas D, Barbotin J-N (1990) Appl Microbiol Biotechnol 33:619
206. Chang HN, Moo-Young M (1988) Appl Microbiol Biotechnol 29:107
207. Chen T-L, Humphrey AE (1988) Biotechnol Letters 10:699
208. Monbouquette HG, Ollis DF (1988) Bio/Technology 6:1076
209. Burrill HN, Bell LE, Greenfield PF, Do DD (1983) Appl Environ Microbiol 46:716
210. Kingdon CFM (1985) Appl Microbiol Biotechnol 21:176
211. McLoughlin AJ, Whooley MA, O'Callaghan JA (1986) Ir J Food Sci Technol 10:127
212. Whooley MA, McLoughlin AJ (1983) J Gen Microbiol 129:989
213. Whooley MA, O'Callaghan JA, McLoughlin AJ (1983) J Gen Microbiol 129:981
214. Whooley MA, McLoughlin AJ (1982) Eur J Appl Microbiol Biotechnol 15:161
215. Burke RM, Upton ME, McLoughlin AJ (1990) Ir J Fd Sci Technol 14:51
216. Upton ME, McLoughlin AJ, Burke R (1989) In: Ghee AH (ed) Trends in food biotechnology; Proceedings of 7th world congress of food science and technology. SIFST, Singapore, p 129
217. McLoughlin AJ, Licken B, Newell M, Hussey C (1987) In: Neyssel OM, van der Meer RR, Luyben KC (eds) Proc 4th European Congress on Biotechnology Elsevier, Amsterdam, vol 1, p 408
218. Park SH, Lee SB, Ryu DD (1981) Biotechnol Bioeng 23:2591
219. Hannoun BJM, Stephanopoulos G (1986) Biotechnol Bioeng 28:829
220. Mussenden PJ, Keshavarz T, Bucke C (1991) J Chem Tech Biotechnol 52:275
221. Tsay SS, To KY (1987) Biotechnol Bioeng 24:297
222. Berry F, Sayadi S, Nasri M, Barbotin JN, Thomas D (1988) Biotechnol Letters 10:619
223. Eikmeier H, Westmeier F, Rehm HJ (1984) Appl Microbiol Biotechnol 19:53
224. O'Riordan A, McHale ML, Gallagher J, McHale AP (1989) Biotechnol Letters 10:735
225. Mak AL, Trevan MD (1988) Enzyme Microb Technol 10:207
226. Deo YM, Costerton JW, Gaucher GM (1983) Can J Microbiol 29:1642
227. Sutar II, Vartak HG, Srinivasan MC, SilvaRaman H (1986) Enzyme Microb Technol 8:632
228. Klein J, Vorlop K-D (1984) Ger Chem Eng 7:233

229. Wang HY, Hettwer DJ (1982) Biotechnol Bioeng 24:1827
230. Draget KI, Ostgaard K, Smidsrød O (1989) Appl Microbiol Biotechnol 31:79
231. Liou JK, Rosseau I (1986) Biotechnol Bioeng 28:1582
232. Chevalier P, de la Noüe J (1988) Enzyme Microb Technol 10:19
233. Wikström P, Szwajcer E, Brodelius P, Nilsson K, Mosbach K (1982) Biotechnol Letters 4:153
234. Adlercreutz P, Holst O, Mattiasson B (1982) Enzyme Microb Technol 4:395
235. Kokufuta E, Shimoshashi M, Nakamura I (1988) Biotechnol Bioeng 31:382
236. Zache G, Rehm H-J, (1989) Appl Microbiol Biotechnol 30:426
237. Kurosawa H, Ishikawa H, Tanaka H (1988) Biotechnol Bioeng 31:183
238. Cheetham PSJ (1984) US Patent No 4 443 538
239. Jirkü V, Turkova J, Krumphanzl V (1980) Biotechnol Letters 2:509
240. Karel SF, Briasco CA, Robertson CR (1987) Ann New York Acad Sci 506:84
241. Tyagi RD, Ghose TK (1982) Biotechnol Bioeng 24:781
242. Veelken M, Pape H (1982) Eur J Appl Microbiol Biotechnol 15:206
243. Doran PM, Bailey JE (1986) Biotechnol Bioeng 28:73
244. Fletcher M (1986) Appl Environ Microbiol 52:672
245. Wingender J, Winkler UK (1984) FEMS Microbiol Lett 21:63
246. Wingender J, Volz S, Winkler UK (1987) Appl Microbiol Biotechnol 27:139
247. Byrne GS, Ward OP (1987) Trans Brit Mycological Soc 89:367
248. Galazzo JL, Bailey JE (1988) Biotechnol Bioeng 33:1283
249. Galazzo JL, Bailey JE (1990) Enzyme Microb Technol 12:162
250. Galazzo JL, Bailey JE (1990) Biotech Bioeng 36:417
251. Ryu DDY, Kim HS, Taguchi H (1984) J Ferm Technol 62:255
252. Monbouquette HG, Sayles GD, Ollis DF (1990) Biotechnol Bioeng 35:609
253. Hilge-Rotmann B, Rehm H-J (1990) Appl Microbiol Biotechnol 33:54
254. El-Sayed A-HMM, Mahmound WM, Coughlin RW (1990) Biotechnol Bioeng 36:346
255. Wang S-D, Wang DIC (1990) Biotechnol Bioeng 36:402
256. Beunink J, Rehm H-J, (1988) Appl Microbiol Biotechnol 29:72
257. Kutney JP, Choi LSL, Hewitt GM, Salisbury PJ, Singh M (1985) Appl Environ Microbiol 49:96
258. Bettman H, Rehm HJ (1984) Appl Microbiol Biotechnol 20:285
259. Keweloh H, Heipieper H-J, Rehm H-J (1989) Appl Microbiol Biotechnol 31:383
260. Keweloh H, Weyrauch G, Rehm H-J (1990) Appl Microbiol Biotechnol 33:66
261. Heipieper H-J, Keweloh H, Rehm H-J (1991) Appl Environ Microbiol 57:1213
262. Svoboda A, Ouředníček P (1990) Current Microbiol 20:335
263. Nečas O, Svoboda A (1985) In: Peberdy JF, Ferenczy L (eds) Fungal protoplasts. Marcel Dekker, Basel p 115
264. Schnabl H, Scheurich P, Zimmermann U (1980) Planta 149:280
265. Linsefors L, Brodelius P (1985) Plant Cell Reports 4:23
266. Mori A, Matsumoto N, Imai C (1989) Biotechnol Letters 11:183
267. Morris P, Fowler MW (1981) Plant Cell Tiss Org Cult 1:15
268. Morris P, Smart NJ, Fowler MW (1983) Plant Cell Tiss Org Cult 2:207
269. Hamilton R, Pedersen H, Chin C-K (1984) Biotechnol Bioeng Symp 14:383
270. Hamer G (1990) In: de Bont JAM, Visser J, Mattiasson B, Tramper J (eds) Physiology of immobilized cells, Elsevier, Amsterdam. p 15

Evaluation of Biomass

A. Singh[1], R.C. Kuhad[2], V. Sahai[1] and P. Ghosh[1]
[1] Department of Biochemical Engineering & Biotechnology, Indian Institute of Technology, Hauz Khas, New Delhi 110016, India
[2] Department of Microbiology, University of Delhi South Campus, Benito Juarez Road, New Delhi 110021, India

Evaluation of biomass concentration is an important problem encountered in many microbial and other bioprocesses. It determines the catalytic activity of the microbial cell in a given time. Various direct and indirect methods for the estimation of biomass have been developed using physical and biochemical techniques. Despite many promising classical methods available, the evaluation of microbial growth in bioprocesses may sometimes become laborious, impracticable and give erroneous values. Various methods for enumeration of organisms and determination of biomass, including recent developments in monitoring biomass concentration for the control of biotechnological processes, are discussed taking into the consideration their practical importance, usefulness and constraints in application.

Advances in Biochemical Engineering/
Biotechnology, Vol. 51
Managing Editor: A. Fiechter
© Springer-Verlag Berlin Heidelberg 1994

1 Introduction

Biomass estimation is one of the most important process variables in microbial and other bioprocesses. Its determination leads to the understanding of the efficiency of a biocatalyst in a bioreaction system. The metabolic behaviour is directly related to the growth of the organism which makes it necessary to obtain a balance sheet comprising nutrients consumed and products formed per unit of cell population. Although biomass concentration is a simple measure for the available biocatalyst, it is a key variable in measuring rates of growth and product synthesis, yield coefficients, and also for the calculation of specific rates and mass balance in any bioprocess. Thus an accurate method for the real-time biomass estimation during a bioprocess is an important goal to be achieved.

Classical methods for biomass determination may be based on cell number or cell mass. Methods dependent on cell number are observational, based on physical and microbial activity. These methods include total and viable counting of cells. Total counts usually do not differentiate between active and dead cell populations whereas a few methods may provide the counts of viable or active cells. However, viable counts do not distinguish between cells and clumps of cells and also they are subjected to conditions very different from those in bioprocesses [1]. Viable counts usually underestimate the microbial community when compared to direct count methods. There is a considerable interest in alternative methods for biomass determination. Indirect methods usually estimate the components of the cells. They do not require visual or cultural examination of organisms and depend upon a specific chemical component that is only present in the cell [2]. Biomass evaluation is further complicated when the insoluble substrates are employed in the medium. In that case conversion from counts to biomass by estimating volumes of the microorganisms can result in large errors [3].

Amongst the many measuring techniques proposed for biomass determination, an ideal measure would be the direct monitoring of biocatalyst in a bioreaction system which is the current approach to the control of industrial bioprocesses. During recent years, various sensors have been developed for on-line monitoring of the biological processes [4, 5]. The electronic sensors transmit low-impedance analog or digital signals that can be readily related to the rate of metabolism or biomass. However, a number of difficulties arise when we use these modern methods. In this contribution, we have attempted to discuss various methods for the evaluation of biomass and new approaches useful in monitoring and control of bioprocesses.

2 Direct Methods for the Evaluation of Biomass

A comprehensive summary of direct methods used for the evaluation of biomass is provided in Table 1.

2.1 Conventional Methods

2.1.1 Cell Mass

Dry weight is a rather widely used method for estimation of biomass as well as in basic calibration of other methods [6]. The weight of the dried solid matter of micro-organism affords a better measure of their protoplasm. The cells from a known volume of culture are washed free from the media components with distilled water. Washed cells are placed in a weighed vessel, dried to a constant weight by freeze drying or heating in an oven at 105 °C, cooled in a dessicator and weighed. Cells may be separated either by centrifugation or by filtration. Centrifugation is the cheaper method but it has certain disadvantages including the requirement of higher amount of biomass and continuation of cellular activities which might result in the increase or decrease of the cell mass. The washing of cells also requires extra care in order to avoid losses of cells. The filtration method is expensive, because of the cost of the filter but it is more rapid and requires smaller sample volumes. Further, drying introduces the errors arising from decomposition of biological material during the dehydration process [7, 8]. Moreover, dry weight cannot be considered as true biomass specifically when insoluble solid substrates are used in process.

Table 1. Direct methods for the evaluation of biomass

Method	Remarks
Cell mass	Simple gravimetric method; drying and weighing time-consuming; errors arising from decomposition of biological materials; not applicable when insoluble substrates are used
Turbidity	Rapid; low sensitivity; calibration problem; particle interference
Viable count	Based on ability of cells to grow; large sampling errors; usually under- or over-estimate the active cells; statistically questionable
Epifluorescence filter technique	Simple and rapid epifluorescence microscopic counting; reproducibility questionable
Fluorescent-antibody technique	Based on immunological properties of cells; sensitive; can be used in situ; not suitable for microbial systems producing slime; species specific
Micro-ELISA	Conjugation of antibodies with enzymes; sensitive; suitable for large molecules; comparatively slow; also detects antigens of non-viable cells
Coulter counter	Rapid; problems in high cell density cultures; dilution problems
Electron microscopy	Claimed to be simple and rapid; expensive instrumentation; limited applications

Microbial biomass is sometimes determined by wet weight. The moist surface growth on a solid medium is scrapped from the medium and weighed at once. In bacteria forming capsules and slime, the wet weight may greatly overestimate the amount of protoplasm, since it includes the weight of these highly hydrated extracellular substances. Estimation of wet weights are not precise because of the problem of evaluating the amount of intracellular water and also the water wetting the cell surfaces [9].

2.1.2 Turbidity

Turbidity measurement is another widely used method for the estimation of cells in suspension. The turbidity is measured in reference to scattered light from a suspension. The turbidity of a suspension can be determined in spectro-photometer by measuring the light lost from the beam by scattering. The expression is similar to Beer's law except that the term extinction coefficient is replaced by a constant called the turbidity coefficient. A standard plot can be made of $\log I_0/I$ against either the total count or the dry weight [10]. The concentration factor applies mainly to protoplasmic mass as the size of the organism as well as their number determines turbidity. However, transferring cells from one medium to another or washing disturbs the osmotic situation in the cells which may change the cell/medium ratio and refractive index and also turbidity without altering the cell count or total mass [10]. The calibration curve applies only to a particular organism grown under a particular set of growth condition. A new curve must be prepared if a change is made in either of these. Further cells grown in high carbohydrate or fat content frequently have a high turbidity.

2.1.3 Viable Count

Viable counting methods are also widely used to estimate cell populations [11]. The enumeration is based on the ability of the microbe present in a sample to grow either in liquid culture media or media gelled with agar or on the surface of membrane filter. Viable counting methods may be further divided into plate count method, membrane filter method and most probable number (MPN) method.

In the plate count method, appropriately diluted samples are poured or spread on agar plates and colonies developed are counted after a certain incubation period. In order to avoid sampling and over-crowding errors, plates containing 50–500 colonies should be counted [12]. Serial 10-fold dilutions of the bacterial suspension are prepared and either poured in or spread onto the surface of nutrient agar in sterile petri dishes. Plates are incubated at an appropriate growth temperature. The number of colonies are counted on the plate and multiplied by dilution factor to obtain viable count per ml in the original sample. When counting anaerobic bacteria it is necessary to treat diluents to remove dissolved air prior to use. It can be done by using non-toxic reducing agents [10]. Pipetting should be done carefully to avoid aeration and if possible anaerobic chambers should be used.

The membrane filter method is based on the use of a highly porous cellulose acetate membrane of different pore sizes which prevents the passage of bacteria, however, large volumes of water can pass through under pressure. Bacteria retained on the membrane surface are incubated at appropriate temperature. Thereafter, the number of colonies are counted. Membrane filters are manufactured by Sartorius GmbH, Germany; Millipore Filter Corporation, USA; and Oxoid Ltd, U.K. Filters with a pore size of 0.43–0.47 µm are usually recommended. The major advantage of membrane filtration over conventional plate count methods is its rapidity and large volumes of sample can be processed where cell population is low since the concentration and inoculation steps are completed in one step [13].

The MPN method is commonly used with highly selective media and provides an estimate of the number of viable cells present in a given sample which are capable of growing in liquid growth medium. In this method, serial 10-fold dilutions of the sample are made and an aliquot of each dilution is transferred to the growth media. It is then incubated and the number of positive tubes (showing turbidity) from each dilution recorded. The degree of dilution of the sample required and the number of positive tubes at different dilutions are used to determine MPN of the microorganism referring to probability tables [11, 14]. The MPN method has been widely used to determine coliform and faecal streptococci in water samples [15] as well as nitrifying and sulphate reducing bacteria [16, 17]. MPN values for a given microbial population can now be calculated using simple computer programmes which provides estimates of MPN for any combination of sample volumes, dilution levels and number of replicates (see Ref. [18] for details). MPN method is particularly useful when low cell numbers are required to be determined. Counts obtained by MPN are usually higher than plate count methods, however the major disadvantage is the large sampling error which can be over-come by increasing the number of replicates or the dilution factor.

2.2 Specialized Techniques

2.2.1 Epifluorescence Filter Technique

The application of epifluorescence microscopy is generally said to be one of the best methods available for counting microorganisms. The method involves staining of microbes with a fluorochrome, collecting cells on a membrane filter and finally counting using epifluorescence microscopy [19–22]. With this procedure, fluorochrome (acridine orange or diamidino-2-phenylindole) is added to the sample for a contact time of a few minutes and then filtered through a polycarbonate membrane. The membrane is rinsed with a volume of sterilized water equal to total sample volume which removes excess stain. Bacteria stained with acridine orange fluoresce green while debris appears either red, orange or yellow whereas diamino-2-phenylindole stained cells fluoresce bright blueish-white and particulate matter stains yellow [23]. The number of bacteria in the sample is calculated from the mean cell number, the volume of the sample filtered, the effective area of filtration and graticule area using the formula [13], $n = yAd/av$, where n is number of cells per ml of sample, y is mean cell count, A is effective area of filtration, d is dilution factor, a is graticule area and v is volume of filtered sample.

The method is simple but its reproducibility must be checked several times before using it for biomass estimation. Measurement of cell counting/concentration also requires that cells are suspended singly and not in aggregates. In the later case, separation of cells by disintegrating aggregates is necessary [24]. A wide range of epifluorescence microscopes are available (Quantimet 800, Cambridge Instruments; Artek 810, Artek System; IBAS System, Zeiss Kontron) and the linkage of an image analysis system to the microscope is a recent development [19, 25].

2.2.2 Fluorescent-Antibody Technique

Fluorescent-antibody technique can be used to enumerate specific groups of microorganisms in situ. The technique is widely used in medical microbiology and pathology but now has been introduced into microbial ecology and environmental biotechnology [26]. The method, in general, is to inject cellular components of bacteria into a laboratory animal. The antibodies produced are purified and tagged with an appropriate fluorochrome such as fluorescein isothiocynate. The antibodies combine with antigenic microbes, if present, and cause them to fluoresce. The number of fluorescing cells are then counted using an epifluorescence microscope. Direct and indirect fluorescent-antibody techniques are employed to determine the number of specific cells in the biofilm of a fixed-bed reactor used for the secondary treatment of waste water [27]. However, direct techniques are not found suitable in microbial systems producing slime which prevents the immune reaction between fluorescing antibodies and the antigenic surface of the specific cells. Indirect technique has been used to count nitrifying, sulphate-reducing and methanogenic bacteria [28, 29]. However, the major problem is the non-specific adsorption of fluorescent antibodies to substrate colloids and films. This problem may be solved by pretreating them with a gelatin-rhodamin isothiocynate conjugate to suppress non-specific antibody adsorption as well as serving as a counterstain [30]. The method is useful in biological studies since it identifies and enumerate microorganisms in a single step. The limitation is that it is species specific and hence to broaden the spectrum of this technique we would have to produce antibodies against a range of similar organisms and tag this with fluorescein isothiocynate as fluorochrome to produce a polyvalent stain.

2.2.3 Micro-ELISA

Application of micro-ELISA technique for sulphate reducing bacteria was developed by Bobowski and Nedwell [31]. Antisera is produced by administering cell free extracts of bacterial culture into rabbits. An antiserum mixture is then prepared in order to detect and count sulphate reducing bacteria. Although, the method is comparatively slow and requires several washing and incubation stages, it is very sensitive and suitable for assaying large molecules. Antibodies are conjugated with enzymes using glutaraldehyde so that the resulting conjugates have both an enzyme activity and an immunological property [32]. This can be further quantified by their ability to degrade a suitable substrate. Horseradish peroxidase and alkaline phosphatase are the commonly known enzymes with o-phenyl diamine and p-nitrophenyl phosphate as substrates, respectively.

Micro-ELISA has not yet become a routine method in most microbiological laboratories. Comparisons of micro-ELISA technique with most probable number methods revealed that counts were greater in the first technique which was probably due to the broader specificity of antibody conjugate and in addition also detects cellular antigens of non-viable cells unlike most probable number method which only detects viable cells [31]. Numbers as low as 10^3 cells per membrane filter could be detected using this technique [32].

2.2.4 Coulter Counter

The Coulter counter manufactured by Coulter Electronics Inc., Florida has been widely used to count microorganisms [33–35]. A suspension of cells is passed through a small orifice that has two electrodes suspended on either side. Resistance caused by the orifice is measured after passing current between two electrodes. The pulse generated by each cell is amplified and recorded electronically, giving a count of the number of cells flowing through the aperture. Cells can be counted in the medium in which they are growing. Cell concentrations must be kept low for accurate results since high cell densities limit the accuracy by coincidence counts when more than one cell occupies the counting orifice [36]. This problem can be solved by diluting the cell suspension, however, saline used as diluent may cause shrinkage of cells due to osmotic stress. Any such changes in cell volume can be prevented by using a phosphate buffer.

2.2.5 Electron Microscopy

In microscopic observations, cells are viewed under a microscope. Examples are enumerating cells using a Petroff-Hausser bacterium counter and a Levy Haemocytometer [9]. However, the resolution of the electron microscope makes it superior to the light microscope for estimation of bacterial size. The use of scanning electron microscopy for counting bacteria on membrane filters has been reported earlier [37, 38], however, its use for biomass estimation could not become common mainly because of the difficulty in producing quantitative preparations. Recently Borsheim et al. [39] described a method for enumeration and biomass estimation of bacteria by transmission electron microscopy. Samples were harvested by ultracentrifugation directly onto the electron microscopy grids (JEOL 100CX TEM) and counted. For counting and sizing unstained grids were used since fixation and staining may lead to the loss of unpredictable amounts of elements [40]. Cell volumes were calculated as $\pi/4 \, W^{-2} (L - W/3)^{-1}$, where W is the width and L is the length of the cell. Dry matter content was determined by X-ray microanalysis. The dry weight/volume ratio for bacteria was 600 fg μm^{-3} of dry weight. The method was claimed to be simple and rapid but it has only been used for estimation of bacterial biomass in natural environments.

3 Indirect Methods for the Evaluation of Biomass

Table 2 summarizes various indirect methods for the evaluation of biomass.

3.1 Cellular Components

The determination of biomass and activity of microbes in the bioprocesses using insoluble substrate presents a complex analytical problem. A number of studies have demonstrated that classical methods of isolating and subsequent culturing

Table 2. Indirect methods for the evaluation of biomass

Method	Remarks
Cellular components	
Nucleic acids	Fluorimetry/spectrophotometry; systematic errors in absorption spectra due to the presence of other nucleotides and peptides; depends on physiological factors
Protein	Simple but laborious; depends on nutritional and physiological factors; questionable validity
Cell carbon/phosphate	Not generally applicable; sampling errors; washing leaches out the sample compounds; underestimate the biomass concentration
Polysaccharides	Errors due to sample preparation; probable side reactions; difficult to calibrate
Lipids	Depends on physiological state; active metabolism; rapid turn-over in dead cells
ATP	Relatively specific; sampling and sample preparation critical; can be used in processes involving insoluble substrates
Metabolic activity	Directly related to growth; very sensitive to physiological changes in cells; overestimation of cell concentration.

of organisms are not adequate for enumerating in such processes. The major problem is the removal of organisms from the solid substrate. In that case, cells may be hidden in substrate particles and the conversion of counts to biomass, by estimating cell volume may result in large errors. Moreover, classical methods provide a limited insight into the metabolic function and activity of the cell. Such problems have stimulated research on the development of indirect methods based on 1. cellular components – biomass is estimated by measuring the concentration of a specific biochemical component and 2. metabolic activity – biomass is estimated by measuring the rate of isotope incorporation from labelled precursors or by vital staining. The major requirements for determining biomass by measuring the concentration of a cellular component are 1. the presence of the component in fairly uniform concentration in the cells and 2. it should be possible to quantitatively extract and analyze with appropriate precision. However, most of these methods are time-consuming and the calibration is difficult with questionable precision. Some common methods for indirect estimation of biomass are discussed below.

3.1.1 Nucleic Acids

Cell biomass is comprised of the sum of many metabolic processes such as nucleic acid and protein biosynthesis and assimilatory process [41]. Thus metabolites such as DNA and RNA contents have been correlated to the extent of growth [42, 43]. Nucleic acids are universally present in all living cells and may be used as a general indicator of microbial biomass. However, it cannot differentiate bacterial biomass from fungal biomass. Methods of nucleic acids determinations are relatively precise but are subject to systematic errors [44, 45]. Cells also contain nucleotides and peptides much smaller than nucleic acids and proteins, and interfere with absorption spectra by giving the same colour reaction. Moreover, biological calibration is required before these parameters can be taken as measures of the growth, since cell composition varies with time and also depends on nutritional and physiological factors. Corrections must also be made for the measured component present in the substrate.

3.1.2 Protein

Measurement of protein content by estimating the amount of nitrogen present in cells have been correlated to the microbial biomass [46, 47]. The cells from a known volume of culture are washed by centrifugation to free them from media components and extracellular excretion products. The total nitrogen content of the cell can be estimated by the micro-Kjeldahl method or using Nessler reagent [48]. The percentage nitrogen thus obtained can be multiplied by a factor of 6.25 to obtain the protein content in the sample. Another method based on protein was suggested by which the extent of substrate degradation and contribution of fungal mycelium can be calculated using an equation, $ax + by = c(x + y)$; where $(x + y)$ is the weight of total biomass recovered, comprised of fungal mycelium (y) and undegraded substrate (x); a is the crude protein content of substrate and b is the crude protein content of fungal mycelium; and c is the crude protein content of total biomass recovered [47]. This method has been used in determining biomass during cellulose biodegradation studies [49, 50]. In the presence of cellulose, fungal biomass can also be determined by extracting the mycelial protein in NaOH solution and measuring the protein. Fungal biomass is calculated assuming that the fungal mycelium contains 40% protein. The method based on protein measurement is simple but laborious. It depends very much on physiological and nutritional factors, also the validity is questionable.

3.1.3 Cell Carbon/Phosphate

Measurement of cell carbon and phosphate have also been used as indices of growth [51–53]. Automated methods are available for the determination of total carbon in biological materials [51]. For phosphate determination cells are separated by centrifugation, washed thoroughly to make it free from media components and digested in the presence of triacid mixture and measured colorimetrically [54]. However, the estimation of these components can be correlated with biomass only when residual media components containing the element in question have been removed. Harvesting and washing involves losses of cells, and the latter leaches out water soluble compounds which leads to underestimation of biomass. These methods are not in common use.

3.1.4 Polysaccharides

Chitin, a polymer of N-acetylglucosamine, is a major component of fungal cell walls. Chitin assay, one of the most common indirect methods for the estimation of fungal biomass, has been applied to a wide range of solid substrates like decaying of wood, food products and cereal grains [57, 58] and in microbial production of gibberellic acid in solid state culture [59]. The chitin assay is based on the acid hydrolysis of the polymer followed by determination of hexosamine [60–62]. This method for biomass determination is relatively insensitive and has been a subject of number of critical appraisals [63]. The principle sources of error are the tendency for the chitin content of the mycelium to vary with age and physiological state of the mycelium thus the conversion factor from glucosamine to fungal biomass cannot be standardized. Other factors such as growth conditions may also cause variation in the chitin concentration. The concentration of chitin ranges between 10 and 250 mg g^{-1} dry weight in various fungal species. The chitin content in $Coriolus\ versicolor$ was $2.4\ \text{mg g}^{-1}$ when grown in wood, whereas in vivo studies revealed $12.4\ \text{mg g}^{-1}$. The presence of extraneous hexosamines within the substrate also cause interference with the assay. Presently chitin assay for determining fungal biomass is only limited to monoculture conditions.

In an analogous manner to the use of chitin determination in fungi, muramic acid may be used to determine bacterial biomass. Most of the prokaryotic cells contain muramic acid (a peptidoglycan) as a component of the cell wall [64]. Several methods have been proposed for the determination of muramic acid in microbial biomass, involving the conversion of muramic acid to lactate followed by enzymatic or chemical analysis of lactate concentration. Components of the bacterial cell wall are recovered after hydrolysis with strong hydrochloric acid and then purified by cation-exchange chromatography [65–67]. Muramic acid is quantified by analysis of lactate released by subsequent alkaline hydrolysis. A more sensitive method is to form a derivative and then analyse muramic acid by GC [68] or by HPLC [69]. The sensitivity of these methods is about 10^{-11} mol of muramic acid which corresponds to about 10^8 bacteria. The muramic acid content in bacterial monocultures does not vary greatly with growth conditions, however, its concentration in Gram-positive and Gram-negative bacteria is significantly different [70].

Muramic acid content of several bacterial isolates has been found to be, on an average 19.4 mg g^{-1} biomass C for Gram-positive bacteria and 7.2 mg g^{-1} for Gram-negative bacteria [65]. An equivalence in biomass estimations between measurements based on muramic acid, other chemical measures and microscopic counts has been demonstrated. However, the method is very laborious, time-consuming and difficult to calibrate. Probable side reactions during sample preparation must also be considered since the individual chains of N-acetylmuramic acid are interconnected by short peptide bridges containing specific amino acids such as D-alanine and m-diaminopimelic acid [71].

3.1.5 Lipids and Their Derivatives

Plasma membranes contain phospholipid as a major component in all living cells but is not a storage product of microorganisms. Lipid phosphate could be correlated to other indicators of biomass such as ATP, thus it has been proposed as an indicator of microbial biomass [72, 73]. The method is based on the extraction of lipid phosphate from biomass with a suitable organic solvent and subsequent phosphate analysis. White [72] has developed a method to determine phospholipid as a biomarker of microbial biomass. The method involves extraction of a known amount of sample with a one-phase chloroform : methanol : phosphate buffer mixture. Further, a two phase system is made by adding equal volumes of chloroform and buffer. The upper methanol water phase is removed and the chloroform layer containing lipids is collected. Chloroform is evaporated by using a rotary evaporator. The phosphate content of the dried lipid sample is determined colorimetrically [54]. Values of phospholipid content in various monocultures have been used to convert phospholipid measures to microbial biomass [74, 75]. However, phospholipid content in microbes varies among different taxa [76, 77]. It has been reported that the phospholipid content of bacteria varies from 4–91 mg g^{-1} dry weight [76]. The equivalence of microbial biomass estimations using phospholipid measurements, ATP analysis, microscopic counts and several other chemical methods have been demonstrated [78], however, monoculture studies have shown that concentration and composition of phospholipids are affected by growth conditions and media composition [79]. Phospholipids are actively metabolized during growth of bacteria and they have relatively rapid turnover in dead cells [80].

Ergosterol (ergosta-5,7,22-trien-3-β-al) the predominant sterol found in most fungi, can form the basis of another method in indirect biomass estimation [81, 82]. It can be measured on the basis of its characteristic ultraviolet absorbance. Ergosterol can be recovered from the sample by extraction with methanol, saponification and reextraction with hexane [83]. Thereafter, the analysis is done by HPLC using a UV detector, taking advantage of the characteristic UV absorption of ergosterol at 282 nm, which differs significantly from that of animal and plant sterols [84]. The physical losses during extraction has been found to be less than 8% by this method. Alternatively, ergosterols can be extracted with chloroform-methanol mixture, purified with silicic acid column, recovered from neutral lipids and analysed by capillary gas chromatography [85]. The ergosterol content of fungal mycelium depends on the species and growth conditions. A range of 2.3–5.9 mg l^{-1} dry weight of ergosterol was obtained when three fungal species were compared by growing under different conditions and times [84]. Fungal ergosterol showed a similar relationship to mycelium growth in liquid culture but the results with solid substrates has been found to be unsatisfactory [85].

Gram-negative bacteria contain unique lipopolysaccharide polymers in the outer cell membrane which can form a basis for the determination of biomass. Lipopolysaccharides consist of a lipid (lipid A) and a core polysaccharide. The most common method for lipopolysaccharide determination is based on the limulus amoebocyte lysate test [86]. This test is rapid and sensitive, with a detection limit of about 0.4 pg of lipopolysaccharide. However, there are several problems associated with this method. Quantitative extraction of lipopolysaccharide from the sample is difficult. The specificity of this test in some cases is questionable [87]. It has also been shown that lipopolysaccharide of dead bacteria is lost rapidly [88]. The method is mostly used to determine biomass in environmental samples.

3.1.6 Adenosine Triphosphate (ATP)

Another promising indirect method is the use of ATP to estimate total active biomass. This method is based on the assumption that active microbial cells possess a fairly constant amount of ATP which can be easily measured by bioluminiscence [46]. The principle of the ATP measurement is its reaction with luciferin catalysed by luciferase enzyme.

$$\text{Luciferin (reduced)} \quad \xrightarrow{\text{Luciferase}} \quad \text{Luciferin (oxidized)}$$
$$+ \text{ATP} + O_2 \qquad\qquad\qquad + \text{AMP} + \text{pp} + \text{light}$$

One photon of light is produced per molecule of ATP hydrolysed which can be measured using a photometer. For details the comprehensive review by Karl [89] may be consulted. Thiery and Chicheportiche [90] described a simple and rapid ATP extraction procedure in which sample is homogenized in dimethyl-sulphoxide and quickly centrifuged. A known volume of preserving buffer (0.75 mM glycine, 4.4 mM Mg-EDTA, pH 7.5) is added, quickly frozen in liquid nitrogen and stored at $-18\,°C$. Extracts are thawed immediately before measurements. Crude firefly luciferase is reconstituted with 20 mM Tris buffer (pH 7.8) and the assay is performed in a photometer tube [91]. The light emission from the reaction can be measured with a spectrophotometer, a liquid scintillation counter or a photometer – like Luminometer. The concentration of ATP in the sample is determined by interpolation from a standard ATP calibration curve. The luciferin-luciferase system, most commonly used for ATP measurement, has a sensitivity of about 10^{-4} mol of ATP [92]. This sensitivity corresponds to approximately 10^3 bacterial cells by assuming ATP content of 4 nmol mg^{-1} dry weight and an average dry weight of bacteria 6.4×10^{11} cells per g [89, 93].

This method has been successfully employed for estimating biomass during biological humification and detecting contamination of food [92]. Intracellular ATP concentration has been found to be well correlated with the fungal growth and specific rate of cellulose consumption as a function of time [94]. The potential source of error in the ATP determination lies in sampling and extraction technique since ATP assay is time dependent. Thus the crucial point of the ATP assay is the necessity of processing the sample rapidly after collection since the physiological state of the cell changes considerably after sampling. However, once the technique is standardized, consistent results can be obtained.

3.2 Metabolic Activities

CO_2 production is another possible index of growth especially during the early stages of cell growth [95]. pCO_2 can be indirectly measured as H^+ ion concentration of a bicarbonate buffer behind a glass permeable membrane. CO_2 reacts with H^+ and HCO_3^- and the resulting pH is measured with an electrode (Ingold AG). CO_2 output is useful when rapid growth and vigorous cell activity are occurring. The rates of evolution of CO_2 could be directly proportional to the cell mass, under given conditions. However, variations in cultural conditions may sometimes pose problems in using the CO_2 evolution method for biomass estimation. Even small changes in pH values may also affect the amount of CO_2 evolved considerably [96]. A radiometric method was developed by measuring

$^{14}CO_2$ evolved from the metabolism of ^{14}C-glucose for the detection of some food-borne bacteria [97]. It requires 8–10 h for the detection of aerobic as well as anaerobic bacteria and has potential for sterility testing of food.

Measurement of O_2 uptake is complementary to CO_2 evolution and is a measure of cell activity [98]. Oxygen is taken up by living microorganisms often at rather constant rates. Therefore, O_2 uptake may be correlated to active biomass, provided major physiological changes like adaptation to a new energy source are not occurring [99]. Volesky et al. [100], while studying aerobic cultivation of various yeast species on different carbon sources, observed that O_2 uptake rate can be used as an indicator of metabolic activity. In aerobic cultures, the rate of heat release correlated well with oxygen cell yield (amount of O_2 uptake per unit of biomass formed) [101].

Active electron transport systems are almost universal indicators of bacterial metabolic activity [102]. The tetrazolium salt, 2-(p-iodophenyl)-3-(p-nitro-phenyl)-5-phenyl tetrazolium chloride (INT) is preferentially reduced by the electron transport chain of respiring bacteria with the deposition of opaque, red formazan crystals. These crystals are sufficiently large and dense to be clearly visible within actively respiring cells by epifluorescence microscopy. Thus actively respiring bacteria can be enumerated by incubating with INT and then counting the proportion of bacteria containing one or more dark formazan crystals [103]. However, one major drawback is that formazan deposits dissolve in the immersion oil used as mountant. The method is generally used in the activated sludge process and not in common practice. Lopez et al. [104] have also reported a method based on the measurement of INT-dehydrogenase for activated sludge process control. INT-dehydrogenase activity was found directly proportional to INT dosage and inversely proportional to biomass concentration but over limited ranges. However, this method is not precise and is generally not applicable.

Fluorogenic esters like fluorescein diacetate, polar non-fluorescent esters, have been used as vital stains to determine metabolically active cells [105]. These esters are hydrolysed in the cell by esterases to the non-polar fluorescein which is retained within the cell. Thus only active organisms with esterase activity will fluoresce under suitable illumination. They can be used successfully with yeast but sometimes give variable results [105]. Metabolically active cells can also be determined by using a combination of autoradiography and microscopy [106]. These methods require enzymatic activity in the presence of substrates and are subject to limitations associated with the density of organisms and thickness of the biofilm. When substrates are introduced to measure activity, high levels of metabolic activity are induced with possible disturbance artifacts [106, 107]. It is also known that enzymatic methods do not work correctly in soil samples, probably enzymes are also inhibited by the toxic substances. Thus for indirect biomass determination CO_2 production rates would be better measures.

Another method to determine the metabolically active cells is based on viability. This is estimated on the basis of the ability of individual cell to catalyse a biochemical reaction. The method is commonly used to characterize yeast and animal cell cultures [108–110]. Cells that are viable contain an enzyme that reduces methylene blue or trypan blue to the respective leuko form, whereas dead cell do not perform this reaction. It is simple, therefore to distinguish between living and dead cells by examining them microscopically. Living cells are unstained and dead cells stained blue. Although the method is rapid, it may sometimes overestimate the viable cells since lysed or lysing dead cells are not accounted.

4 Methods for On-line Evaluation of Biomass

During recent years, significant developments have been made on optimum control methods and their applications in bioprocesses both at laboratory and industrial levels [111–115]. Rapid reliable and quantitative methods for on-line determination of biomass have been developed during the last decade (Table 3). Although a number of difficulties arise when we use modern methods to optimize a bioprocess, the results of many researches in this area clearly show that the application of these methods can bring substantial improvements in existing processes and the productivities by real time analysis [116–122].

4.1 Microcalorimetry

A close relationship between entropy, free energy and enthalpy and the growth of baker's yeast was observed earlier [123]. Therefore, calorimetric data for heat released may relate well to the microbial growth. Heat output depends on the

Table 3. Methods for the on-line evaluation of biomass

Method	Remarks
Microcalorimetry	Rapid; based on heat evolution; well correlated to biomass concentration and O_2 uptake; also suitable for slow growing cells like hybridoma cells; disturbance in activity due to temperature fluctuation; suitable for bioprocesses containing insoluble substrates.
Fluorescence	Specific for NAD(P)H, F350, F420 etc.; depends on physiological state of cell
Electrical properties	Electrical conductance/impedance; may be rapid and accurate; signals depend on media composition and pH; bubble interferences
Automated methods	Sensitive; interferences by particulate matters or gas bubbles; reference problem
Spectroscopy	Rapid; selective; calibration problem; depends on composition and physiological state of cells

velocity with which cells derive their metabolism and may represent a valuable means of monitoring microbial processes [124]. The method based on calorimetry measures of overall metabolic capacity which is dependent on the number of cells and microcalorimeters like the LKB Bioactivity Monitor, LKB Flow Microcalorimeter and BRIC Microcalorimeter have been used for this purpose.

A certain amount of heat is always evolved during the course of a biological process, whether the system is aerobic or anaerobic or whether the final product is metabolite or biomass [125]. There have been many microcalorimetric studies of yeast growth where the total heat evolved closely follows the curve for biomass production [126–134]. Cooney et al. [129] used a simple thermistor technique for measuring the rate of heat production during fungal cultivation by monitoring the media temperature increase. This method has been used to measure the rate of heat evolution during the process of novobiocin synthesis by *Streptomyces niveus*, cellulase synthesis by *Trichoderma viride* [130] and for the growth of *Saccharomyces cerevisiae* [131].

Ciba-Geigy Heat-Flux Calorimeters (BSC-81) have been used by Marison et al. [132] for bioprocess monitoring. They observed a good correlation between heat and biomass production when *Kluyveromyces fragilis* was grown in an aerobic culture on a defined lactose medium. The ratio between heat and biomass production of 10.59 kJ g^{-1} of cell dry weight remained constant despite the change of growth rate during several experiments. Heat-Flux calorimetry could play an important role in monitoring and controlling growth of microorganism. Good correlations between microbial growth, the heat evolution and oxygen uptake using dynamic calorimetry was also demonstrated earlier [133, 134]. These correlations demonstrate the usefulness of calorimetry in monitoring industrial bioprocesses particularly for continuous measurement of biomass concentration. Although dynamic calorimetric techniques possess an advantage in their relative simplicity, they require continuous attention without computer control. Temperature fluctuations may also significantly disturb the metabolic activity particularly at the end of the exponential growth phase when rapid metabolic changes occur. Only limited success has been achieved in designing instruments capable of continuous on-line monitoring of biomass concentration [134]. Nevertheless, the measurement of the rate of heat evolution by calorimetry may be useful in the quantitative assessment of total active biomass formation in the bioprocesses containing insoluble substrates [130, 135, 136]. Calorimetry may also be an excellent tool for the determination of slow growing organisms like hybridoma cells [137].

4.2 Culture Fluorescence

It has been observed that the measurement of culture fluorescence could provide a real-time estimate for the biomass concentration under controlled process conditions [138]. Major fluorescence results from the intracellular NADH and NADPH. However, when the environmental conditions are variable in the system, the fluorescence data may be sensitive for biomass concentration since the intracellular NADH and NADPH levels change very rapidly under given conditions [138, 139]. Fluoresensors have been developed to determine the intracellular pool of NADH and NADPH. In many cases fluorosensor signals have been related to the physiological state of the population and not the biomass [140–145]. There are several reports on using fluorosensor probe, e.g. Fluoromeasure Detector System (BioChem Technology, Malvern), for the determination of biomass [146–149]. However, the major constraint is the change in physiological state. Thus any change in the physiological state of the cell may result in the specific change of NAD(P)H pool. Sometimes signals cannot be validated when the sensitivity of the sensor is low and background fluorescence is too high.

Methanogenic bacteria contain unique fluorophores which give a measurable fluorescence at certain wavelengths. The fluorescent properties of methanogens are due to the presence of coenzymes F_{420} and F_{350}. This unique property has been used to enumerate methanogenic bacteria in their natural environment [150, 151]. Recently fluorosensor signals have been used in monitoring methanogenic process controls, using probes designed to measure coenzyme F_{420} [152, 153]. However, these methods have less application in biomass determination since cellular F_{420} varies greatly.

4.3 Electrical Properties

The presence of metabolising microorganisms can also be detected by monitoring changes in the electrical conductance of culture media [154]. The method is based on the principle of conductance, the ability to carry electrical current. The ionic concentration of the medium is changed by the metabolites of the multiplying organism. Conductance, which is dependent on the ionic concentration of the solution can, therefore, be a reliable measure of the cell population. The method based on this technique was found rapid and more accurate and can detect even a single viable microorganism [155]. The process is now automated and can be of considerable sophistication.

A 'Biomass Monitor' (Aber Instruments Ltd., Aberystwyth, U.K.), devised with an in situ, steam sterilizable capacitance probe was used to follow biomass concentration on-line in bioreactors from 20–2000 litre total volume [156]. A linear correlation was found between on-line capacitance measurement and off-line measurements (OD_{620}, dry cell weight and colony forming units) for biomass concentration up to $30 \, g l^{-1}$ (*Streptomyces virginiae*), $106 \, g l^{-1}$ (*Saccharomyces cerevisiae*) and $89 \, g l^{-1}$ (*Pichia pastoris*). The on-line measurement was slightly influenced by variation in agitation speed and strong extraneous radio frequencies. It has also been used to monitor microbial growth on solid substrates [157] and to study shearing sensitivity of plant cells in suspension culture [158]. However, with this monitor, the capacitance could be measured only when the conductance of the media was less than 20 mS [156].

The signal which depends on biomass also depends on other factors such as media composition and pH and is also affected by bubble interferences [159, 160]. Measurements of conductivity using Digital Conductivity Meter, PW9527 (Philips Co., England) have been found to be linearly correlated with biomass concentration of plant cells up to $10 \, g \, l^{-1}$ [160].

Microbial growth and activity has also been determined by measuring impedance change in the growth medium caused by the products of bacterial and yeast metabolism [161, 162]. Yeast growing in wort causes an increase in impedance, whereas bacteria cause a decrease. This difference may also be diagnostic in quality control. The forcing test used in brewing industry for the detection of spoilage microorganisms in beer could be shortened from a few weeks to a few days using impedimetry (Bactometer 32, Bactomatic Inc., Marlow) [161]. Measurements based on conductance/impedance have been correlated to the growth of yeast in bioprocesses [162] and also for detecting and determining wine spoiling yeast [163]. Direct electrochemical measurements have been found slow and unreliable [164, 165]. Recently Ding and Schmid [166] developed an electrochemical measurement in a flow injection system for on-line monitoring of E. coli. The method is rapid and well correlated to classical method of colony forming unit determination.

It is well known that dielectric properties of ionic solutions are affected by the presence of cellular materials. Harris et al. [167] have recently pointed out that the radiofrequency dielectric properties of microbial cell suspensions are direct and a monotonic function of the radius and volume fraction of the particle constituting the suspended phase. They significantly differ from those of particulate matter, gas bubbles and aqueous solutions. Measurement of these variables permit the direct estimation of biomass during microbial processes, in situ and in real time [168]. An instrument (Bugmeter Viable Cell Counter, Aber Instrument Ltd., Aberystwyth) based on this technique has been commercialized and may be used on-line. Non-cellular particulate matter were not found to interfere significantly in biomass estimation. Limitations of this method for measuring biomass have recently been pointed out by Sonnleitner et al. [146]. The maximally acceptable conductivity can be easily achieved in intermediately concentrated media, but suffers from low sensitivity in dilute media and fatal interferences by the media conductivity in concentrated media.

4.4 Automated Methods

Sensors based on optical density (OD) are now available which work on the principle of transmission or scattering of light. A fairly good correlation between OD and biomass concentration has been observed for different microorganisms by using external or in situ OD sensors (Sybron-Brinkmann; LT 201, Komatsugawa/Biolafitte; MEX-3 OD Sensor, Bonnier Technology Group) [169–171]. Different sensors are significantly different with respect to sensitivity. Interference by particulate matter other than cells or gas bubbles may result in erratic results. Problems in high density cultures were solved by Nielsen et al. [172]

who obtained reliable signals by using flow injection analysis (FIA) and steady-state absorbance measurement. FIA has also been applied to estimate microbial biomass directly [172] or indirectly [173]. A control loop for the measurement of dissolved oxygen concentration (< 100 ppb) based on a fast but non-sterilizable sensor (Marubishi DY-2) has been devised [174], and Wilson [175] investigated on-line biomass monitoring based on the dynamic oxygen balancing.

Component balancing by applying computers and software sensors are also useful in many cases [176]. Hardware signals when converted into the useful units, may cause different ranges of resolution depending on the magnitude of the signal. Locher et al. [177] have shown that even common conversions, e.g. correlation of biomass with the OD of a culture or with the fluorescence sensor signal, may be limited to the condition met during evaluation and can be misleading. Strässle et al. [178, 179] have shown a structured and/or segregated mathematical model for the proliferating biosystem, S. cerevisiae, which can be verified by reliable experimental measurements. This leads to successful progress in getting dynamic insight into cells and entire populations [180–183]. Biomass can reasonably be calculated using a verified model. However, the major drawback is that the analyses of basic variables are not sufficiently available to calculate software sensor. Further there are a series of assumptions and simplifications which do not often hold good [146].

Automatic filtration devices with stainless steel support filter screens (Gelman 78781) have been developed to measure the concentration of biomass in large samples [184, 185]. These filters are best suited for fungal cultures. In this method, flux of filtrate and filter cake build-up is monitored after passing the sample through the filter. Biomass concentration in the range of 2–40 g l^{-1} can be measured using this method [184]. Appropriate descriptive models are used to calculate biomass concentration. Density measurement with the help of an Acoustic Resonance Densitometer (Paar DMA-55/DMA-60, Austria) was found linearly correlated to the optical density in the range of 0–500 OD units [186–187] and hence formed the basis of another method for biomass estimation. Control values of the cell free medium must also be obtained in order to avoid temporal variations of the media density during cultivation which may otherwise give erratic results. This method is also successfully employed in mammalian cell culture in which up to 6×10^7 cells per ml can be measured [188].

4.5 Spectroscopy

The metabolic state of microorganisms can be measured by short wavelength near infrared spectroscopy [189]. It rapidly measures several hundreds of spectra and this technique can be applied to measure biomass concentration. Similarly measurements of protein by infrared-photoacoustic spectroscopy may be used to indirectly determine biomass concentration in different microbial cultures [190].

Recently Manoharan et al. [191] reported that bacteria and bacterial endospores can be potentially detected and identified by UV resonance Raman spectroscopy. Their results indicate that UV-absorbing bacterial taxonomic markers can be selectively excited to give rise to characteristic resonance Raman bacterial fingerprints which is a potential method for rapid identification. From the same spectra, the concentration of cell wall components can be indirectly correlated to the biomass. However, the difference in cell composition with cultural conditions and varying stages of cell development make the use of spectra complicated for identification. Nevertheless, this is a highly sensitive method because of the negligibly low protein fluorescence at < 250 nm.

5 Conclusion

A very common problem in the fields of Microbiology and Biotechnology is to determine the concentration of cells present in a sample. Classical methods may be dependent upon cell mass or number. Alternative methods for estimating biomass by measuring cellular components have also been developed. However, individual methods suffer from various drawbacks including: 1. a substantial processing time, 2. lack of sensitivity, and 3. lack of amenability to continuous monitoring of the biomass. Most of the methods are not suitable for industrial applications because of the interfering characteristics of the bioprocesses e.g. the presence of insoluble substrates, coloured medium or no-specificity of the method. During recent years, various measuring techniques have been developed for on-line monitoring of the biological processes. Direct measurement of process characteristics is of interest. However, despite the significant developments, there is no ideal sensor for monitoring biomass. Since they have still not reached the stage of general application, in many cases we must content ourselves with indirect measurements. Efforts made during the last few years clearly indicate that substantial improvement has been made in this area. The development of an ideal biomass sensor to determine biological activities and intracellular components is based on the needs and constraints of the industrial bioprocesses.

Acknowledgement. Valuable suggestions and information provided by Prof. A. Fiechter and Dr. P.K.R. Kumar are gratefully acknowledged.

6 References

1. Olsen RA, Bakken IR (1987) Microb Ecol 13: 59
2. White DC (1986) Arch Hydrobiol 31: 1
3. Paul EA, Ladd JN (eds) (1981) Soil biochemistry, vol 5, Marcel Dekker, New York, p 415

 4. Wang HY (1984) Biotechnol Bioeng Symp 14: 601
 5. Graham A, Moo-Young M (1985) Biotechnol Adv 3: 209
 6. Roberts RB, Abelson PH, Cowie DB, Bolton ET, Britten RJ (1955) Studies on the biosynthesis in *Escherichia coli*. Carnegie Institute Pub. 607, Washington
 7. Powell EO (1963) J Sci Food Agric 14: 1
 8. Hobson PN, Mann S (1970) Automation, mechanization and data handling in microbiology. Academic, London
 9. Pirt SJ (1975) Principles of Microbe and Cell Cultivation, Blackwell Scientific, Oxford
10. Collee JG, Duguid JP, Fraser AG, Marrison BP (1980) Practical medical microbiology, 13th edn. Churchill Livingstone, London
11. Meyenell GG, Meyenell E (1970) Theory and Practice in Experimental Bacteriology. Cambridge University Press, Cambridge
12. Cruickshank R, Duguid DM, Chih-Hua W (1975) Medical microbiology, 12th edn. Churchill Livingstone, Edinburgh
13. Jones JG (1979) A guide to methods of estimating microbial numbers and biomass in fresh water. Biological Association, Windermere
14. Taylor J (1962) J Appl Bacteriol 25: 54
15. American Public Health Association (1976) Standard methods of the examination of water and wastewater, 14 edn. Washington
16. Macfarlane GT, Herbert RA (1984) J Gen Microbiol 130: 2301
17. Battersby NA, Stewart, DJ, Sharma AP (1985) J Appl Bacteriol 58: 425
18. Clarke KR, Owens NJP (1983) J Microbiol Meth 1: 133
19. Austin B (ed) (1988) Methods in aquatic microbiology. John Wiley, Chichester, p 27
20. Hobbie JE, Daley RJ, Jasper S (1977) Appl Environ Microbiol 33: 1225
21. Zimmerman R (1977) In: Rheinheimer G (ed) Microbiol ecology of a brackish water environment. Springer, Berlin Heidelberg, New York, p 103
22. Daley RJ (1979) In: Costerton JW, Colwell RR (eds) Native aquatic bacteria: Enumeration, activity and ecology. American Society of Testing and Materials, Philadelphia
23. Wynn-Williams DD (1985) Soil Biol Biochem 17: 739
24. Fry JC (1990) Meth Microbiol 22: 41
25. Björnsen PK (1978) Appl Environ Microbiol 36: 584
26. Schmid EL (1973) Bull Ecol Res Comm 17: 67
27. Szwerinski H, Gaiser S, Dardtke D (1985) Appl Microbiol Biotechnol 21: 125
28. Belser LW, Schmid EL (1978) Appl Environ Microbiol 36: 584
29. Smith AD (1988) Arch Microbiol 133: 118
30. Bohlool BB, Schmid EL (1973) Bull Ecol Res Comm 17: 336
31. Bobowski S, Nedwell DB (1987) In: Hopton JW, Hill EC, Industrial microbiological testing, Blackwell Scientific, Oxford, p 171
32. Engvall E, Perlman P (1971) Immunochemistry 8: 871
33. Evans JH, McGill SM (1969) Hydrobiologia 35: 401
34. Sheldon RW, Parsons TR (1967) A practical manual on the use of the Coulter counter in marine science. Coulter Electronics, Canada Ltd
35. Sieburth JM (1979) Sea microbes. Oxford University Press, New York
36. Kubitschek HE (1969) Meth Microbiol 1: 593
37. Krambeck C, Krambeck H-J, Overbeck J (1981) Appl Environ Microbiol 42: 142
38. Watson SW, Novisky JJ, Quinby HL, Valois FW (1977) Appl Environ Microbiol 33: 940
39. Borsheim M, Bratbak G, Heldal M (1990) Appl Environ Microbiol 56: 352
40. Heldal M, Norland S, Tumyr O (1985) Appl Environ Microbiol 50: 1251
41. Moo-Young M, Moreira AR, Tengerdy RP (1983) The filamentous fungi, vol 4. Edward Arnold, London, p 117
42. Moreira AR, Phillips JA, Humphrey AE (1978) Biotechnol Bioeng 21: 1501
43. Holme NA, McIntyre AD (1977) Methods for the study of the marine benthos. Blackwell Scientific, Oxford
44. Chattopadhyay NC, Nandi B (1977) Phytopath Z 89: 256
45. Peach K, Tracey MV (1955) Modern methods of plant analysis. Springer, Berlin Heidelberg New York, p 246
46. Wang DIC, Cooney CL, Demain AL, Dunnill P, Humphrey AE, Lilly MD (1979) Fermentation and enzyme technology. John Wiley, New York
47. Garg SK, Neelkantan S (1982) Biotechnol Bioeng 24: 2407
48. Lang CA (1958) Anal Chem 30: 1692

49. Singh A, Abidi AB, Darmwal NS, Agrawal AK (1988) MIRCEN J Appl Microbiol Biotechnol 4: 473
50. Singh A, Abidi AB, Darmwal NS, Agrawal AK (1988) Biol Mem 14: 53
51. McDonald AMG (1963) Ind Chem 39: 265
52. Benett EO, Williams RP (1957) Appl Microbiol 5: 14
53. Hosler P, Johnson MJ (1953) Ind Eng Chem 45: 871
54. Galnous DS, Kapoulos A (1966) Anal Chim Acta 34: 360
55. Swift MJ (1973) Soil Biol Biochem 5: 321
56. Swift MJ (1973) Bull Ecol Res Comm 17: 323
57. Chen GC, Johnson BR (1982) Appl Environ Microbiol 46: 13
58. Hicks RE, Newell SY (1984) Oikos 42: 355
59. Kumar PKR, Lonsane BK (1987) Biotechnol Bioeng 34: 276
60. Frankland JC, Lindley DK, Swift MJ (1978) Soil Biol Biochem 10: 323
61. Ride JP, Drysdale RB (1971) Physiol Plant Pathol 1: 409
62. Ride JP, Drysdale RB (1972) Physiol Plant Pathol 2: 7
63. Sharma PO, Fisher PJ, Webster JP (1977) Trans Br Mycol Soc 69: 479
64. King JD, White DC (1977) Appl Environ Microbiol 33: 777
65. Miller WN, Casida LE (1970) Can J Microbiol 16: 299
66. Casergrande DJ, Park K (1978) Soil Sci 125: 181
67. Gunnarsson T, Tunlid A (1986) Soil Biol Biochem 18: 595
68. Tunlid A, Odham G (1983) J Microbiol Meth 1: 63
69. Moriarty DWJ (1983) J Microbiol Meth 1: 111
70. Ellwood, DC, Tempest DW (1972) Adv Microb Physiol 7: 83
71. Schleifer KH, Kandler O (1972) Bacteriol Rev 36: 407
72. White DC, Davies WM, Nickels JA, King JD, Bobbie RJ (1979) Oecologia 40: 51
73. Slater JH, Whittenbury EJ, Wimphery JNT (eds) (1983) Microbes in their natural environments. Society of General Microbiology, p 37
74. Gehron MI, White DC (1983) J Microbiol Meth 1: 23
75. Blakwill DL, Leach FR, Wilson, JT, McNabb JF, White DC (1988) Microb Ecol 16: 73
76. Kates M (1964) Adv Lipid Res 2: 17
77. Wossef MK (1977) Adv Lipid Res 15: 159
78. Kowalenko CG, McKercher RB (1970) Soil Biol Biochem 2: 269
79. Lechevalier MP (1977) Crit Rev Microbiol 7: 109
80. White DC, Tucker AN (1969) J Lipid Res 10: 220
81. Nannipieri P, Johnson RL, Paul EA (1978) Soil Biol Biochem 10: 223
82. Logal DM (1988) In: Ratledge C, Wilkinson SG (eds) Microbial lipids, vol 7. Academic, London, p 699
83. Grant WD, West AW (1986) J Microbiol Meth 6: 47
84. Nes WR (1977) Lipid Res 15: 233
85. Matcham SE, Jordan BR, Wood DA (1985) Appl Microbiol Biotechnol 21: 108
86. Ford SR, Webster JJ, Leach FR (1985) Soil Biol Biochem 17: 811
87. Suzuki M, Mikami T, Matsumoto T, Suzuki S (1977) Microbiol Immunol 21: 419
88. Saddler JN, Wardlow AC (1980) Antonie von Leeuwenhoek J Microbiol 46: 27
89. Karl DM (1980) Microbiol Rev 44: 739
90. Thiery A, Chicheportiche R (1988) Appl Microbiol Biotechnol 28: 199
91. Deming JW, Picciolo GL, Chappelle EW (1979) In: Costeron JW, Colwell RR (eds) Native aquatic bacteria: Enumeration, activity and ecology. American Society of Testing and Materials, Philadelphia, p 88
92. Karl DM, Holm-Hanson O (1976) Anal Biochem 75: 100
93. Gray TRG, Hissel R, Duxbury T (1974) Rev Ecol Biol Soil 11: 15
94. Cochet N, Tyagi RD, Ghose TK, Lebeault JM (1984) Biotechnol Lett 6: 155
95. Kavanagh F (ed) (1963) Analytical microbiology, Academic, New York
96. Calam CT (1969) Meth Microbiol 1: 567
97. Previt JJ (1972) Appl Microbiol 24: 535
98. Harris D (1979) In: Grossbard E (ed) Straw decay and its effects on utilization and disposal. John Wiley, Chichester, p 265
99. Siegmund D, Diekman H (1989) Appl Microbiol Biotechnol 32: 32
100. Volesky B, Yerushalmi L, Luong JHT (1982) J Chem Technol Biotechnol 32: 650
101. Luong JHT, Yerushalmi L, Volesky B (1983) Enzyme Microb Technol 5: 291
102. Zimmerman R, Itturiaga R, Backer-Birck J (1978) Appl Environ Microbiol 36: 926

103. Dutton RJ, Bitton G, Koopman B (1983) Appl Environ Microbiol 46: 1263
104. Lopez JM, Koopman B, Bitton G (1986) Biotechnol Bioeng 28: 1080
105. Patton AM, Jones SM (1975) J Appl Bacteriol 38: 199
106. Ramsay AJ (1984) Soil Biol Biochem 16: 475
107. Baath E (1988) Soil Biol Biochem 20: 123
108. Postgate JR (1969) Meth Microbiol 11: 611
109. Painting K, Kirsop B (1990) World J Microbiol Biotechnol 6: 346
110. Combrier E, Matezean P, Ronot X, Gachelin H, Adolphe M (1989) Cytotechnol 2: 27
111. Sonnleitner B, Fiechter A (1989) GBF Monogr 13: 75
112. Locher G, Sonnleitner B, Fiechter A (1990) Biopr. Eng. 5: 181
113. Locher G, Sonnleitner B, Fiechter A (1991) J Biotechnol 19: 127
114. Sonnleitner B, Locher G, Fiechter A (1991) J Biotechnol 19: 1
115. Locher G, Sonnleitner B, Fiechter A (1992) J Biotechnol 25: 23
116. Picque D, Corrieu G (1988) Biotechnol Bioeng 31: 19
117. Schügerl K, Lübbert A, Scheper T (1987) Chem Ing-Tech 59: 701
118. Rohner M, Locher G, Sonnleitner B, Fiechter A (1989) J Biotechnol 9: 11
119. Arnold MA, Ostler TJ (1988) Crit Rev Anal Chem 20: 149
120. Schügerl K (1991) Analytische Methoden in der Biotechnologie. Vieweg, Braunschweig
121. Münch T, Sonnleitner B, Fiechter A (1992) J Biotechnol 22: 329
122. Münch T, Sonnleitner B, Fiechter A (1992) J Biotechnol 24: 299
123. Battley EH (1960) Physiol Plant 13: 628
124. Boe I, Loverien R (1990) Biotechnol Bioeng 35: 1
125. Eriksson R, Holme J (1977) Flow microcalorimetry applied to microbial processes. LKB Application Note No. 267, LKB Produkter AB, Stockholm
126. Shaarachmidt B, Lamprecht I (1976) Experientia 32: 1230
127. Lamprecht I (1980) Growth and metabolism in yeast. In: Beezer AE (ed) Biological microcalorimetry. Academic, London, p 43
128. Miles RJ, Beezer AE, Perry BF (1987) Growth and metabolism of yeast. In: James AM (ed) Thermal and energetic studies of cellular biological systems. John Wright, Bristol, p 106
129. Cooney CL, Wang DIC, Mateles RJ (1968) Biotechnol Bioeng 11: 269
130. Mou D-G, Cooney CL (1976) Biotechnol Bioeng 18: 1371
131. Wang H, Wang DIC, Cooney CL (1978) Eur J Appl Microbiol Biotechnol 5: 207
132. Marison IW, Biron B, von Stockar U (1985) Thermochim Acta 85: 493
133. Volesky B, Luong HT, Thambimuthu KB (1978) Can J Chem Eng 56: 534
134. Luong JHT, Volesky B (1982) Can J Chem Eng 60: 163
135. Fardeau M-L, Plasse F, Belaich JP (1980) Eur J Appl Microbiol Biotechnol 10: 133
136. Gustafsson K, Gustafsson L (1985) J Microbiol Meth 4: 103
137. von Stockar U, Marison IW, Birou B (1988) On-line calorimetry for process control. In: 1st Swiss-Japanese Joint Meeting on Bioprocess Development, Interlaken, Switzerland
138. Zabriskie DW, Humphrey AE (1978) Appl Environ Microbiol 35: 337
139. Scheper T, Lorenz T, Schmid W, Schügerl K (1986) J Biotechnol 3: 231
140. Meyer HP, Beyeler W, Fiechter A (1984) J Biotechnol 1: 341
141. Armiger WB, Forro JF, Montalavo LM, Lee JF (1989) Chem Eng Comm 45: 197
142. Leist C, Meyer HP, Fiechter A (1986) J Biotechnol 4: 235
143. Reardon KF, Scheper T, Bailey JE (1987) Biotechnol prog 3: 153
144. Siano SA, Muthrasan R (1991) Biotechnol Bioeng 37: 141
145. Walker CC, Dhurjati P (1989) Biotechnol Bioeng 33: 500
146. Sonnleitner B, Locher G, Fiechter A (1992) J Biotechnol 25: 5
147. Taya M, Yoshikawa M, Kobayashi T (1989) J Chem Eng Japan 22: 89
148. Beyeler W, Einsele A, Fiechter A (1981) Eur J Appl Microbiol Biotechnol 13: 10
149. Müller W, Wehnert G, Scheper T (1988) Anal Chim Acta 213: 47
150. van Bruggen JJA, Stum CK, Vogels GD (1983) Arch Microbiol 136: 89
151. van Bruggen JJA, Stum CK, Zwart KB, Vogels GD (1985) FEMS Microb Ecol 31: 187
152. Peck MW, Chynoweth DP (1990) Biotechnol Lett 10: 17
153. Peck MW, Chynoweth DP (1992) Biotechnol Bioeng 39: 1151
154. Richards JCS, Jason AC, Hobbs G, Gibson DM, Christie RH (1978) J Phys 11: 560
155. Hagen D (1990) Proc Biochem 25: 4
156. Fehrenbach R, Comberbach M, Petre JO (1992) J Biotechnol 23: 303
157. Davey CL, Penaloza W, Kell DB, Hedger JN (1991) World J Microbiol Biotechnol 7: 248
158. Markx GH, Ten Hoopen HJG, Meijer JJ, Vinke KL (1991) J Biotechnol 19: 145

159. Connolly P, Lewis SJ, Corry JEL (1988) J Food Microbiol 7: 3
160. Taya M, Hegglin M, Prenosil JE, Bourne JR (1989) Enzyme Microb Technol 11: 170
161. Evans HAV (1982) J Appl Bacteriol 53: 423
162. Ebina Y, Ekida M, Hoshimoto H (1989) Biotechnol Bioeng 33: 1290
163. Henschkke PA, Thomas DS (1988) J Appl Bacteriol 64: 123
164. Harris CM, Kell DB (1985) Biosensors 1: 17
165. Sakoto K, Tanaka H, Samejima H (1981) Ann NY Acad Sci 369: 321
166. Ding T, Schmid RD (1990) Anal Chim Acta 234: 237
167. Harris CM, Todd RW, Bungard SJ, Lovitt RW, Morris JG, Kell DB (1987) Enzyme Microb Technol 9: 181
168. Kell D, Markx GH, Davey CL, Todd RW (1990) Trend Anal Chem 9: 190
169. Hong K, Tanner RD, Malaney GW, Wilson DJ (1987) Proc Biochem 22: 149
170. Geppert G, Thielemann H, Langkopf G (1989) Acta Biotechnol 9: 541
171. Iijima S, Yamashita S, Matsunaga K, Miura H, Morikawa M (1987) J Chem Technol Biotechnol 40: 203
172. Nielsen J, Nikolajsen K, Benthia S, Villadsen J (1990) Anal Chim Acta 237: 165
173. Valero F, Lafuente J, Poch M, Sola C (1990) Appl Biochem Biotechnol 24: 591
174. Heinzle E, Moes J, Griot M, Sandmeier E, Dunn IJ, Bucher R (1986) Ann NY Acad Sci 469: 178
175. Wilson PDG (1987) Biotechnol Tech 1: 151
176. Roels JA (1980) Biotechnol Bioeng 27: 2457
177. Locher G, Sonnleitner B, Fiechter A (1992) J Biotechnol 25: 55
178. Strässle C, Sonnleitner B, Fiechter A (1988) J Biotechnol 7: 299
179. Strässle C, Sonnleitner B, Fiechter A (1989) J Biotechnol 9: 191
180. Park SH, Hong KJ, Lee JH, Bae JC (1983) Eur J Appl Microbiol Biotechnol 17: 168
181. Sonnleitner B (1991) Bioproc Eng 6: 187
182. Sonnleitner B, Fiechter A (1992) Adv Biochem Eng/Biotechnol 46: 143
183. Chattaway T, Demain AL, Stephanopoulos G (1992) Biotechnol Prog 8: 81
184. Thomas DC, Chittur VK, Cagney JW, Lim HC (1985) Biotechnol Bioeng 27: 729
185. Reuss M, Boelcke C, Lenz R, Peckman U (1987) Biotech Forum 4: 3
186. Blake-Coleman BC, Clarke DJ, Calder MR, Moody SC (1986) Biotechnol Bioeng 28: 1241
187. Clarke DJ, Blake-Coleman BC, Carr RJG, Calder MR, Atkinson T (1986) Trend Biotechnol 4: 173
188. Kilburn DG, Fitzpatrick P, Blake-Coleman BC, Clarke DJ, Griffiths JB (1989) Biotechnol Bioeng 33: 1379
189. Cavinato AG, Ge Z, Mayes DM, Callis JB (1990) A biomass sensor based on visible and short wavelength near infrared spectroscopy, 3rd International Symposium Analytical Methods in Biotechnology, San Francisco
190. Gordon SH, Greene RV, Freer SN, James C (1990) Biotechnol Appl Biochem 12: 1
191. Manoharan R, Ghiamati E, Dalterio RA, Britton KA, Nelson WH, Sperry JF (1990) J Microbiol Meth 11: 1

Convective Drying of Bacteria

II. Factors Influencing Survival

L. C. Lievense[1] and K. van 't Riet[2]
Food and Bioprocess Engineering Group, Wageningen Agricultural University,
P.O. Box 8129, 6700 EV Wageningen, The Netherlands

In the previous part of this review, the parameters of the drying process that can be important for the survival of bacteria upon drying, were reviewed. In this part the other factors which can be important for survival, will be discussed. The discussion starts with the mechanisms that can be responsible for thermal and dehydration inactivation. Moreover, the influence of storage conditions on the stability of dried bacterial cultures will be reviewed.

---

[1] Present address: Unilever Research Laboratory, P.O. Box 114, 3130 AC Vlaardingen, The Netherlands
[2] To whom correspondence should be addressed

Advances in Biochemical Engineering/
Biotechnology, Vol. 51
Managing Editor: A. Fiechter
© Springer-Verlag Berlin Heidelberg 1994

1 Inactivation During Drying

To minimize thermal and dehydration inactivation it is important to understand the physiological mechanisms involved in these inactivation processes. In the literature, there is extensive speculation concerning these mechanisms but also well devised theories, based on experimental evidence, exist.

1.1 Thermal Inactivation Mechanisms

Allwood and Russel [1], Corry [2], Tomlins and Ordal [3] and, more recently, Gould [4] have written reviews on the mechanisms of thermal inactivation of vegetative bacterial cells. From these reviews it is clear that thermal inactivation is still not fully understood. It is generally assumed that there are four main primary sites for thermal damage that can cause injury or death. These sites are 1) DNA, 2) RNA (including ribosomes (rRNA)), 3) proteins (enzymes), and 4) the cell membrane. When the cell wall is also added [1], most essential bacterial cell components are named. It is clear that most, if not all cell components can be subjected to thermal damage. It is therefore extremely difficult to defect a single key factor that causes cell injury or death and most likely, a multiplicity of effects is involved. DNA [4] and (r)RNA [1] are considered the most probable targets for heat inactivation.

The mechanism of heat damage under normal 'wet' conditions is not well understood and it is even less apparent why micro-organisms are generally increasingly heat resistant with decreasing water concentrations [2]. A simple explanation would be that water is involved as a reactant in the heat inactivation. Lowering the water concentration then leads to decreasing heat inactivation rates. At intermediate water concentrations, however, increasing heat inactivation rates are also found at decreasing water concentrations [5]. This is probably due to a transition between different reaction mechanisms involved in the inactivation process.

1.2 Dehydration Inactivation Mechanisms

Mechanisms of dehydration inactivation are summarized in reviews on freeze-drying and yeast drying (mentioned in [6]). It is not always clear if inactivation during convective drying is due solely to dehydration mechanisms, because thermal inactivation can occur simultaneously. Also, during freeze-drying two processes, freezing and drying, occur both of which can influence the inactivation state. As with thermal inactivation, dehydration inactivation may affect a number of different cellular components — these are summarized below.

DNA/RNA. DNA and/or RNA breakdown during (freeze-)drying was mentioned by Wagmann [7], Heckly [8, 9], Webb [10] and Ashwood-Smith [11]. Beker et al. [12, 13] observed chromatin condensation in the early stages of yeast drying and stated that this was a protection mechanism of yeast, and typical for populations with high viability after drying.

Proteins. Scott [14] suggested a reaction between amino groups on cell proteins and carbonyl compounds in the cell during drying and storage. Morichi [15] mentioned that reactive carbonyl compounds, such as ribose, diacetyl and pyruvate, should be removed from the cell suspension before freeze-drying, again because these compounds may react together with amino groups, on essential cell components. Mitic [16] measured a change in amino acid composition of cell proteins and an increased ammonia concentration after freeze-drying of *Lactobacillus bulgaricus*. Mitic concluded that deamination and decarboxylation are important inactivation mechanisms. It is probable that such reactions are the result of severe denaturation of proteins.

The structure of proteins is partially dependent on water, which can form an integral part with the protein [17, 18]. Usually the water which interacts with proteins (and membranes, see below) is referred to as 'bound' water. Many authors suggested a dehydration inactivation mechanism based on the removal of 'bound' water from proteins [13, 19–21]. While the existence of 'bound' water is debatable [22, 23] there is no doubt that the conformation of proteins is destabilized by dehydration, which can result in protein denaturation.

Cytoplasmatic membrane. Cytoplasmatic membrane damage is generally considered as the main mechanism of dehydration damage. Several authors have reported leakage of cellular components (cations, nucleotides, enzymes, proteins, amino acids, or 'UV-adsorbing materials') out of (freeze-)dried cells during rehydration [7, 9, 10, 13, 24]. It can be expected that damage to the cell membrane will lead to leakage of intracellular components. Yet it appears that the release of UV-adsorbing materials is not quantitatively related to inactivation. Brennan et al. [24] reported that membrane damage after freeze-drying of *Lactobacillus acidophilus* was not severe enough to release significant amounts of β-galactosidase, while the release of other UV-adsorbing materials was observed. Wagman [7] found a significantly higher release of UV-adsorbing materials from freeze-dried *Escherichia coli* and *Serratia marcescens* cells when compared to undried cells. Release was strongly dependent on the rehydration medium used but it was not related to the measured survival rate. Webb [10] reported that the release of UV-adsorbing materials from *E. coli* cells with a high survival rate (71%) after freeze-drying, was three times higher than the release from cells with a low survival rate (1%). Morichi et al. [25] found that the materials released from rehydrated freeze-dried *L. bulgaricus* mainly consisted of RNA-related substances but also in this work, no correlation was found between survival and the extend of leakage.

A well devised theory about membrane damage due to dehydration is proposed and reviewed by Crowe et al. [26, 27] and summarized below. Mixtures of phospholipids and water are polymorphic. Even for purified homogeneous phospholipids, there are several structural organization possible when hydrated. The lamellar bilayer structure, in gel or liquid crystalline phase, and the hexagonal phase are the most frequently observed [28]. Compared to the liquid crystalline phase, in the gel phase the molecules are packed more tightly and the acyl chains of the phospholipids are highly ordered, which results in a greater bilayer thickness. The organization of the phospholipid-water system is mainly dependent on composition, temperature and hydration state of the bilayer, and ionic strength

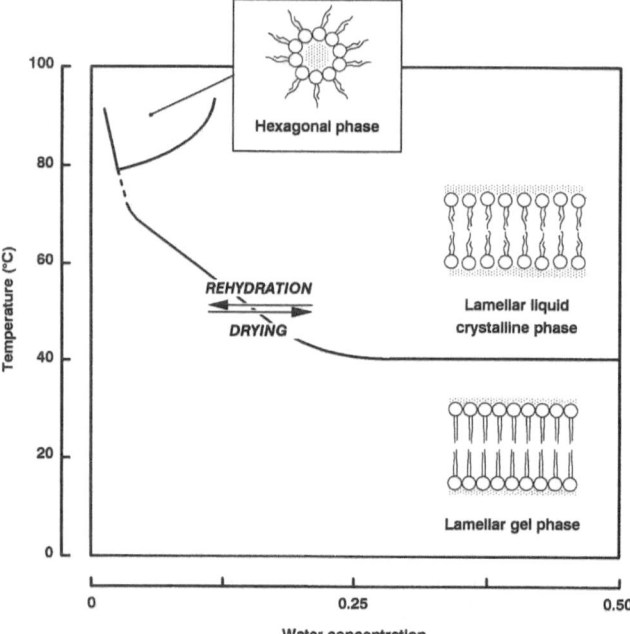

Fig.1. A simplified phase diagram of a single component phospholipid in water. The *solid lines* are the phase transition lines. The *horizontal arrows* indicate the drying and rehydration process. (Axes values are indicative for dipalmitoyl phosphatidylcholine in water [28]) Water concentration (weight of water per weight of solids) kg kg^{-1}.

and pH of the surrounding medium. In these membrane structures, the (bound) water interacts with the phosphate in the phospholipid head group. Figure 1 gives a simplified phase diagram for a one-component phospholipid in water.

The liquid crystalline phase is considered as being representative for the bulk of the phospholipids in the biological membrane. During drying (or rehydration), the phospholipid-water system passes the phase boundary between the liquid crystalline and the gel phase. During drying, the hexagonal phase is usually not reached because it is located in the high temperature region (> 80 °C). The passing of the phase boundary can result in damage to the bilayer due to two mechanisms [26]. Firstly, due to inhomogeneities in the dehydration or heating, the phospholipids in a bilayer undergo a phase transition at different times and this can cause damage to the bilayer. Secondly, biological membranes consist of a mixture of phospholipids and because each phospholipid type enters the gel phase at a different temperature/hydration state and thus different time, this will lead to segretation of the different phospholipids during drying. This segregation is called 'lateral phase separation' and is considered as the most important mechanism in the damaging of biological membranes during dehydration.

Experimental evidence for the lateral phase separation mechanism, carried out with liposome model system consisting of one or several phospholipid types, is given by Crowe et al. [29]. Some results obtained with micro-organisms are in agreement with the membrane phase transition theory. Souzu et al. [30, 31] reported similar changes in the cytoplasmatic membrane of *E. coli* during freezing and freeze-drying. Beker and Rapoport [13] reported mesosome-like formations in the periplasmatic space after rehydration of dried yeast. They assumed that this was related to the segregative function of the cytoplasmatic and other membranes, which were separating damaged structures out of the cells. However, this could also be due to lateral phase separations in the yeast membrane.

Cell wall. Bullock and Lightbrown [32] mentioned alteration in the cell wall of *Streptococcus lactis.* Josic [21] presented electron micrographs (EM) on which cell wall damage and even cell wall separation from the yeast protoplast was shown. Also Sobzack [33] observed morphological changes in the cell wall of dried yeasts. Brennan et al. [24] observed cracks in the cell wall of *L. acidophilus* after freeze and vacuum drying. This author reported that these cracks could be due to the EM preparation, but they were not observed with undried cells.

Other mechanisms. Rogers [34] was probably the first to propose that the inactivation of bacteria during drying was caused by the increasing acid and solids concentration during drying. Obviously, high acid concentrations will be harmful whereas high concentrations of other components will increase the osmotic pressure in the surrounding medium, which can lead to plasmolysis of the cells.

Oxydation reactions are also mentioned as a possible inactivation mechanism. A correlation between the residual activity of dried yeast and the oxidation level of glutathione was reported by Josic [21]. Low residual activities ($<10\%$) occurred at high oxidation levels ($>20\%$). However, there is not sufficient experimental evidence to state that oxidation reactions do occur during drying.

1.3 Conclusions

In conclusion it can be stated that the two mechanisms, thermal inactivation and dehydration inactivation, are being studied but are not yet fully understood. Every cell component can be influenced by high temperatures or low water concentrations. After drying, alterations of the cell membrane are nearly always observed. It might be that damage to the cell membrane is the key factor in the dehydration inactivation. It is even more uncertain if there is also a key factor for thermal inactivation.

2 Factors Influencing the Survival

As was discussed in Part I of this review [6] a number of parameters of the drying process itself can influence the inactivation of bacteria. Nevertheless it is important to realize that other factors can be even more important. A number of these factors are summarized below.

2.1 Bacterial Species

Species differ in sensitivity to drying. Among vegetative bacteria, one can roughly distinguish between sensitive and less sensitive cells, by distinguishing between Gram-negative and Gram-positive bacteria, respectively [2, 8, 9]. Gram-positive cocci (*Streptococcus pyogenes, Staphylococcus aureus*) seem to be the most resistant against (freeze-)drying [8]. Unfortunately, in the production of starter cultures the bacterial species cannot readily be replaced.

2.2 Growth Conditions

The growth phase of the bacterial cells in important for the survival. Probably, the composition of the cell membrane will change with growth conditions, which can influence the membrane phase transition behaviour as described above. Generally, the highest dehydration resistance is found for bacteria harvested in the stationary growth phase. It is not clear at which point in the stationary growth phase the bacterial cells are most resistant. Foster [35], Morichi [15] and Yankov and Brankova [36] found the highest survival of freeze-dried lactic acid bacteria when the bacteria were harvested in the early stationary growth phase. Sapp and Hedrick [37] reported that no significant differences in acid producing activity after spray drying were found for 12, 16 or 24 h old cultures of lactic acid bacteria, while the activity of an 8 h old culture was 10% less. Clement and Rossi [38] and Divies et al. [39] expressed the yeast growth stage as the number of buds present on the yeast cells. Bud forming will be maximal in the exponential growth phase. Clement and Rossi advised, for optimum survival, less than 1% yeasts cells with buds. Divies et al. found, after fluidized bed drying of yeast immobilized in alginate, a survival of 25% for yeast cells in an early stationary growth phase (25% buds) compared to 90% survival for yeast cells in a late stationary growth phase (3% buds). Krallish et al. [40, 41] found that the survival rate of yeast during drying was influenced by the energy levels in the cells. Yeast cells in the stationary growth phase showed an increase in ATP-level when they were dried, whereas with exponential phase cells this behaviour was not observed. The accumulated ATP can serve as an energy source during rehydration of the cells which could result in a higher survival rate. Zikmanis et al. [42] reported similar conclusions.

It was found that trehalose, synthesized by yeast cells during growth, increased the survival after drying. This observation could again be related to the membrane phase transition theory (see Sect. 2.3). During starvation of the yeast cells in the stationary growth phase, an increase in the intracellular trehalose concentration (up to 15% of yeast dry weight) was observed by Beker and Rapoport [13] and Pearce et al. [103]. Gadd et al. [43] reported a negligible dehydration resistance of yeast cells (survival < 10%) in the exponential growth phase, even when large quantities of external trehalose were added. In contrast, survival of yeast cells in the stationary growth phase could be stimulated by addition of external trehalose.

They stated that the protective effect was probably based on a co-operative interaction of internal and external trehalose with the yeast membrane.

Influence of growth on the composition of the bacterial membrane was extensively studied by Zikmanis et al. [44–46]. It is likely that Zikmanis' findings are also related to the membrane phase transition theory of Crowe et al. [26, 27]. Zikmanis et al. reported that the degree of saturation of the fatty acid part of the phospholipid molecules in the membrane could be changed by changing the aeration, temperature, nutrient and growth time. A high temperature and low aeration resulted in a high degree of saturation. Zikmanis et al. [44, 45] reported a strong correlation between saturation and survival (a high degree of saturation (>96% saturated fatty acids) gave high viabilities (>80%) after drying) but in later work [46] this correlation seems to be obsolete.

Other influences of the composition of growth medium were also reported. Beker and Rapoport [13] found that yeast grown on a rich molasses medium survived drying better (90% survival) than yeast grown on synthetic medium (20–40% survival). Wright and Klaenhammer [47] reported that inclusion of calcium in the growth medium favoured the survival of L. bulgaricus after freeze-drying (19% vs. 0.4% without Ca^{2+}), probably by interaction of the Ca^{2+}-ions with the bacterial membrane.

2.3 Protective Additives

Research on the influence of protective additives is abundant. It was stated that one of the reasons that additives have been studied so extensively is that the experimental design is simple and that positive results are readily obtained [8]. This is certainly true, but later [9, 20] it was acknowledged that the use of protective additives is the most fruitful strategy for obtaining optimal survival after (freeze-)drying. Positive influences on survival have been reported for sugars, polyalcohols, carboxylic acid, glycerol, milk and skimmilk, culture medium, polymers (PEG, dextran), proteins, amino acids and salts.

Orndorff and MacKenzie [48], Kilara et al. [49], and Valdez et al. [50–52] tried a variety of additives on freeze drying lactic acid bacteria. Each additive was better than water alone. In view of such wide variety, Orndorff and MacKenzie [48] suggested that there could not be a specific interaction between the additive and the bacterial cell. They stated that an amorphous matrix formed by the additive could be responsible for dispersion of harmful components (such as acids) that leaked out of the cells. Another mechanism proposed by Heckly [8], is simply based on the retention of water by the additive (particularly sugars and poly-alcohols) without a specific interaction. However, without interaction, the additives cannot induce a change in water equilibrium properties of the cells. If so, the amount of water held by the cells per se will be the same at a certain vapour pressure, independent of the presence of additives, i.e. the retained water will be held by the additive and not by the cells [53]. In contrast, other authors [15, 50, 54] state that there is an interaction. Their hypothesis is that the protective effect is based on specific interaction of the protective additive with hydroxilic groups on the cell membrane and proteins. Such interactions (hydrogen bond formation) stabilize these structures by replacing the role of 'bound' water. A similar mechanism was supposed by Beker et al. [12, 13] who mentioned that divalent cations (Ca^{2+}) form salt bridges between phospholipids thus stabilizing the membrane structure.

Studies of the mechanism of protection of proteins and especially membranes, by sugars and divalent cations are reviewed extensively by Crowe et al. [26, 27]. Crowe et al. [29] have shown that retention of water is not required to preserve the structure of dried liposomes. Crowe et al. [27, 55] and Lee et al. [56] report that the stabilizing effect of trehalose, along with other mono-, di- and trisaccharides, is based on direct hydrogen bonding interaction with the phosphate of the phospholipid heads. This interaction lowers the transition temperature between the gel and liquid crystalline phase (see Fig. 2) due to an increase in headgroup spacing and a decrease in the Van der Waals interactions between the acyl chains of the phospholipids. The lower transition temperature permits the membrane to exist in the liquid crystalline phase under conditions at which it normally would be in the gel phase. Therefore, injurious membrane phase transitions during drying and rehydration are avoided. It was suggested that the ability of sugars to stabilize phospholipid bilayers is dependent on the stereo-conformation of the hydroxy groups [29].

In yeast drying, lipophilic substances are added to the compressed yeast. These are called wetting or stabilizing agents [13, 33, 57, 58]. Generally, these substances

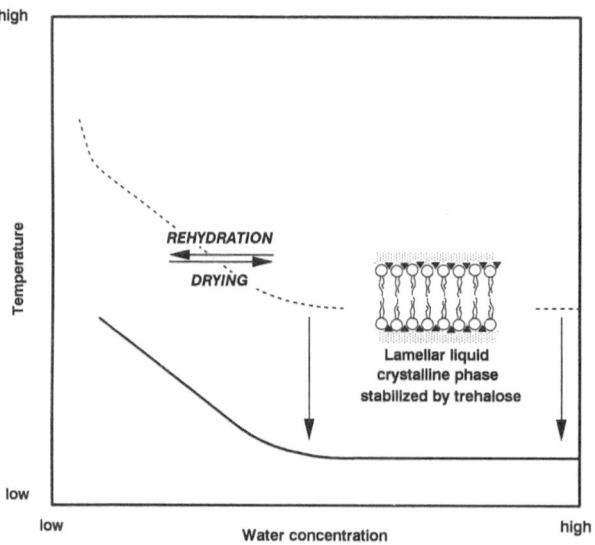

Fig. 2. A simplified phase diagram of a single component phospholipid in water stabilized by trehalose. The *dashed line* indicates the phase transition between the lamellar liquid crystalline and the lamellar gel phase without stabilisation (see Fig. 1). The *vertical arrows* denote the lowering of the transition temperature by trehalose

consist of saturated fatty acid esters of glycerol, sorbitol or polypropylene glycol. Some authors say that mechanical protection is given by these additives [33, 58] because they form a film around the yeast cells. Sobczak illustrated this with EM and again, it is most likely, that an interaction of the additives with the yeast membrane is responsible for the protection.

Possible oxidative reactions during drying [21] can be avoided by the addition of antioxidants. Splittstoesser and Foster [53] found that a combination of thiourea and ascorbic acid was effective in protecting *S. marcescens*, during air drying of suspended cultures. For *S. lactis*, however, this effect was not found. Morichi [15] reported that sodium glutamate protects several lactic acid bacteria against the lethal effect of oxygen during freeze-drying. Similar findings, obtained with freeze and spray drying experiments, were reported by Porubcan and Sellars [59, 60]. Porubcan and Sellars used a combination of ascorbic acid and glutamate or aspartate. De Paz et al. [61] did not observe any effect of ascorbic acid on survival after freeze-drying *S. lactis*. Like oxidation reactions, harmful carbonyl-amino reactions are also thought to be inhibited by addition of an excess of reactive groups. Accordingly, adding amino groups (glutamate) was suggested for neutralizing the carboxyl compounds [20].

2.4 Cell Concentration and pH

It was reported that higher cell concentrations in the suspension to be freeze-dried, generally exhibit higher survival. Kilara et al. [49] found a considerable influence of cell concentration for several lactobacilli. Morichi [25] advised that cell concentrations should be above 10^{10} cells per ml. A possible explanation for this effect could be the release of intracellular components from damaged cells that could protect other cells [8]. It is probable that these components act as antioxidants [9] and if so, the cell concentration could be even more important when no other additives are added. Bozoglu et al. [62] explained the influence of cell concentration with a 'shielding effect' of the bacteria against severe conditions (increasing salt concentration) of the external medium. In this explanation, dead bacteria reduce the interfacial area between viable cells and the external medium.

Several authors have reported an influence of the pH before drying. Schweigart [63] found optimal pH values, different for freeze-, spray and oven drying of *Lactobacillus delbruckii*, all in the range 4.6–5.3. Porubcan and Sellars [59] reported a higher survival for lactic acid bacteria when the pH was set from 4.5 to 6.0–6.5 before freeze-drying (88×10^8 vs 73×10^8 colony forming units per g dried product; initial counts were not reported). Sapp and Hedrick [37] reported a reverse effect at spray drying of a lactic acid bacterium, where neutralization of the acidic culture with sodium or calcium hydroxide resulted in a lower residual acid producing activity than without neutralization. Conversely, Rogers [34] found no effect of neutralization on spray drying of a lactic acid culture. Heckly [9] found that the survival of freeze-dried *S. marcescens* was dependent on the rate at which the pH was adjusted after growth. In view of the contradicting experimental results concerning the influence of cell concentration and pH, no general conclusions can be made. If, however, a low pH reduces the survival after drying, then it is important that no additives are used which can be metabolized to (lactic or acetic) acid before or during the drying process. If this does occur, then addition of a buffering compound can limit the loss of survival.

2.5 Drying Gas, Drying Rate, and Drying Extent

The composition of the drying gas can influence survival. For example, nitrogen may be preferred when oxidation of cellular components plays an important role. There is little research concerning this phenomenon, but Splittstoesser and Foster [53] did indeed find that the survival of *S. marcescens* increased (from 5% to 40%) when nitrogen was used instead of air. Freeman et al. [64] reported that, during spray drying of several bacterial species, the use of nitrogen gave equal or superior survival compared to air. Due to the large amounts of drying gas used, the replacement of air is restricted only to laboratory experiments. Closed-cycle systems [65], where the moistened drying gas can be re-used after condensation of the water, may be economically feasible but, generally, oxidation reactions have to be avoided by the use of antioxidants.

Although little experimental evidence regarding the effects of drying rate on inactivation of bacteria exists, there is much speculation. One can expect that if bacteria are exposed to adverse conditions (high temperatures) or detrimental reactions (oxidation, acid production), then a high drying rate which shortens the drying process if favourable [66–68]. A side effect of an initially high drying rate can be the formation of a dry shell at the surface of the drying particle, which will decrease the overall drying rate. This will render the moisture-temperature history of the cells inside the particle uncontrollable [21].

Only a few authors have discussed the influence of the drying rate itself. Krallish et al. [40] found an increase in the NADH-dehydrogenase level in yeast upon drying. High NADH-dehydrogenase levels can result in higher ATP concentrations in the cells which can enhance recovery during rehydration [41]. Furthermore, it was reported by Marino et al. [69] and Pearce et al. [103], that yeast cells could synthesize trehalose not only during unfavourable growth conditions, but also during drying (from 2.7 to 7% of yeast dry weight, for wet and dried yeast, respectively). Of course, neither the NADH-dehydrogenase concentration nor the trehalose concentration can be increased immediately. This may contribute to the usually low survival rate found when yeast is spray dried. Bullock and Lightbrown [32] and Zimmermann [70] believed that inactivation of bacteria during drying was probably due to the rapid removal of water. Kuts and Tutova [71] mentioned that the rate of drying must be regarded as one of the most important factors that influences survival. They carried out fluidized-bed drying experiments with *Rhizobium pisum* and reported that a two-fold decrease in drying rate in the constant rate period gave a five-fold higher survival (25 vs 5%). Because the initial drying rate was varied by initial water concentration, the moisture-temperature history of the cells could also have influenced the survival, however, drying temperatures were not reported. Nevertheless, the authors concluded from this experiment that for drying of bacterial cells, high drying rates methods have to be avoided. Splittstoesser and Foster [53] did not observe a significant influence of drying rate during drying of *S. marcescens*, but this rate was varied in a narrow range by changing the air flow. Lievense et al. [72] have shown that large differences in drying rate, achieved by drying above saturated salt solutions, in a fluidized bed and in a laboratory spray dryer (characteristic drying times [72] varied from several seconds to several hours), did not affect the inactivation of *Lactobacillus plantarum*. At lower temperatures ($< 50\,^\circ$C) the residual activity was only determined by the water concentration reached. However, in view of these sparse results from isolated experiments, no general conclusion about the influence of the drying rate itself can be drawn.

The survival of bacteria is undoubtedly related to the final water concentration, due to the dehydration inactivation. Obviously, this relation will also depend on the use of protective additives. For example, Splittstoesser and Foster [53] found that when a mixture of dextrin, ascorbic acid, thiourea and ammonium chloride (Naylor-Smith solution) was used during drying of *S. marcescens*, differences in final water concentration between 0.12 and 0.04 kg of water per kg of solids did not affect the survival. However, without this protective additive the influence of final water concentration was significant. A low water concentration is necessary to obtain storage stability and, therefore, an optimum has to be found based on data of survival after drying and stability during storage.

Josic [21] and Kuts and Tutova [71] mentioned that the water concentration in the lower inflection point of the sorption isotherm of the preparation (for yeast: ± 0.07 kg kg^{-1} at $30\,^\circ$C [21]) can be regarded as the lowest possible water concentration for the drying of bacterial cells. This statement is based on the assumption that below this point the structurally 'bound' water is removed which causes severe damage to cell proteins and cell membrane. For *L. plantarum* immobilized in potato starch it was shown that dehydration inactivation started at a water concentration of as low as 0.35 kg kg^{-1} [72]. The lower inflection point of the corresponding sorption isotherm is located significantly below this value (at ± 0.11 kg kg^{-1}). The water concentration at which the inactivation started was obtained by drying above saturated salt solutions and was also derived from fluidized-bed drying experiments [72]. In the latter calculation the water concentration profile inside the particle was taken into account.

Finally, because of the existence of water concentration profiles inside the drying particle, accurate final water concentrations of the cells can be achieved only by controlling the humidity of the drying air. By adjusting this humidity, the final water concentration at the surface and inside the particle can be controlled.

2.6 Rehydration

Rehydration conditions, such as composition, osmolarity, temperature and pH of the rehydration medium and the rehydration rate can significantly influence survival. Some authors regard rehydration as the most important step [73].

Damage to the cell induced by drying does not necessarily lead to cell death. Some bacteria may die, others are reversibly injured and the remainder are apparently unaffected. The injured bacteria may fail to grow on media for growth of intact cells but, these bacteria may be capable of recovering fully if a suitable environments is provided [74]. Undoubtedly, the composition of the rehydration and/or recovery medium can favour the repair of injured cells. Repair of freeze-dried *Salmonella anatum* was independent of pH between 6 and 7 and did not require synthesis of proteins, RNA or cell wall components but did require synthesis of ATP [75]. Hambleton [76] mentioned the requirement of an energy source in the recovery medium and Valdez et al. [77] reported that the recovery of injured freeze-dried lactic-acid bacteria cells was dependent on the plating medium used after rehydration. They found that the highest survival could be obtained on a simple minimal medium, because components in complex media could interfere with repair mechanisms in the cell. Later Valdez et al. [78] showed that addition of Mg^{2+} or Mn^{2+} to the plating medium had a strong stimulating effect on the survival of most of the lactic acid bacteria tested. Hambleton [76] found that di- and trivalent cations in the recovery media strongly reduced the sensitivity of air dried *E. coli* to lysozyme. Morichi [15] found that addition of Mg^{2+} to the rehydration medium decreased the leakage of cellular components from the freeze-dried *L. bulgaricus*. These observations correlate with the interactions of di- and trivalent cations with the cellular membrane, found in other publications [12, 13, 26, 47, 78].

The importance of the temperature of the rehydration medium has been investigated by several authors. Morichi et al. [25] found that 20–25 °C was the optimum rehydration temperature for freeze-dried *L. bulgaricus*. Valdez et al. [79] reported similar results for several other species of lactic acid bacteria. Beker and Rapoport [13] stated that during rehydration of dried yeast the temperature (35–45 °C) should be above the optimum growth temperature (30 °C). It is likely, that a positive effect of higher temperatures is related to the membrane phase transition theory. Indeed, during rehydration, there will be a lower possibility of membrane phase transition at high temperatures than at low temperatures (see Fig. 1). In accordance with this hypothesis, Mahmoud et al. [80] showed a decreasing leakage of cellular components from dried yeast with increasing rehydration temperature (4–40 °C). However, Morichi et al. [25] found that leakage of cellular components from the freeze-dried *L. bulgaricus* increased with increased rehydration temperature (4–50 °C), especially above 20 °C. Moreover, they reported that higher rehydration temperatures (37–50 °C) were detrimental to freeze-dried bacteria but favourable to spray dried bacteria. Dissimilarity in membrane damage, due to the freezing process, could be responsible for this. Heckly [8, 9] stated that there could be a significant difference between species of bacteria with respect to their optimum rehydration temperature.

Heckly [8] reviewed the effect of cell concentration in the rehydration medium on the survival of freeze-dried micro-organisms and stated that high cell concentrations were favourable. Valdez et al. [73] found similar results for different species of lactic acid bacteria. A release of cellular components from severely injured cells may well benefit the repair of other less damaged cells (see also Sect. 2.4).

Quantitative studies on the rehydration rate were not reported but it is assumed that 'gentle' or 'slow' rehydration is beneficial for dried cells [13]. Generally, rehydration of the bacterial cells in moist air (vapour rehydration), compared to simple addition of water, increased the survival of several bacterial species. During rehydration of yeast, this effect was more pronounced at low temperatures (4 °C) than at high temperatures (37 °C) [8]. Also in these experiments, an explanation for the observed behaviour could be related to the phase in which the membrane exists.

2.7 Survival Measurement

Since injured cells can be repaired, the time interval between rehydration, plating on recovery medium and determination of the survival rate can influence the number of viable cells measured [73]. For example, bacterial activity (i.e. lactic acid production rate), measured directly after rehydration, can give lower survival numbers than plate counts [61, 81]. Injured cells can be repaired during plate incubation but not during the activity test. It must be noted that measurement of survival with the plate count method can be problematic. Most starter culture bacteria occur in pairs or even in chains. For a proper application of the plate-count method, the bacteria in these chains have to be separated, otherwise, a dissociation of the chains before, during or after drying will lead to an over-estimation of the survival rate.

3 Stability During Storage

A lot of research is directed towards the storage stability of dried bacteria. There is no doubt that storage stability increases with decreasing temperature [8, 34, 35, 37, 61, 62, 67, 82–86]. Storage of bacterial cultures above 10 °C for several months without a significant loss of activity seems as yet impossible.

Many authors reported a maximum water concentration below which the cells have to be stored. Generally, the maximum concentrations is in the range of $0.01–0.03 \text{ kg kg}^{-1}$ [15, 20, 87] for (freeze-)dried bacteria and in the range of $0.04–0.09 \text{ kg kg}^{-1}$ for yeast [13, 21, 88–91]. Other publications reported that no influences were found in a certain water concentration range [35, 68, 86]. Heckly [8] advised that the lowest water concentration possible should be used, however, extremely low water concentrations lead to a low survival directly after drying. Bousfield and MacKenzie [20] stated that a low inactivation rate during storage is more important than a high viability after drying but in fact, the optimum combination of these two factors has to be determined. In contrast to bacteria in culture collections, this is particularly true for commercial bacterial starter cultures.

The presence of oxygen in the storage atmosphere is detrimental. All authors who investigated this factor found that storage in nitrogen or a vacuum was superior to storage in air [8, 9, 15, 20, 34, 61, 62, 67, 68, 89, 92, 93]. Occasionally, differences in stability were found when the bacteria were stored in a vacuum, in nitrogen or other oxygen-free gases. Unfortunately, flushing times, vacuum pressures and, more preferably, residual oxygen levels are seldom reported. Therefore, differences in stability could have been easily caused by traces of oxygen, since very small amounts of oxygen can cause significant damage [92]. Heckly [92] and Ashwood-Smith [11] reviewed the influences of oxygen on dried bacterial cultures. In the former review it was concluded that the inactivation during storage was related to the formation of radicals in the presence of oxygen. As possible radical reactions, fatty acid oxidation and (further) DNA damage are mentioned [20]. For example, Beker and Rapoport [13] reported that free fatty acids were released during storage of a dried yeast preparation. In contrast, Korobkina et al. [94] reported that, although the stability of a freeze-dried culture of L. acidophilus during storage was higher when compared to a spray dried culture, the fatty acid oxidation rate was also considerably higher. Furthermore, even without oxygen or in the presence of antioxidants, inactivation takes place during storage. It can be concluded that a number of mechanisms of inactivation during storage are staged in literature but still it remains unclear, exactly which of those are relevant for given conditions.

During storage it is important that the packaging material is impermeable to both moisture and oxygen. It is advisable to use polyethylene with aluminium foil [13, 94, 95] because polyethylene alone is found to be moisture permeable [90]. Heckly [8] also expects light to be of influence and advises storage in darkness. Other authors reported higher stability when particles were coated with carboxy-methyl cellulose or alginate [96] or gelatine [82].

Additives can play an important role during storage. Antioxidants can reduce the detrimental effect of oxygen [15, 61] but trehalose and other sugars can also limit the production of radicals [92]. Porubcan and Sellars [59, 60] patented a stabilizing mixture which consists of a combination of ascorbate with glutamate or aspartate. It has been reported frequently, that different additives can be effective for different bacterial species or during different storage conditions [49, 97–99]. Gehrman and Porubcan [100] patented a stabilizing method for freeze-dried L. acidophilus with sun flower seed oil in which no loss in activity was found during storage (120 d at 23 °C) of the oil-suspended dried cells. Other vegetable oils were less effective (cotton seed, corn; 30 and 80% loss, respectively). It can be concluded that additives can be effective in improving the storage stability as well as in reducing the influence of oxygen.

Of interest is the work on stability of intermediate moisture (food) products, reported and reviewed by Slade and Levine [101]. This work is summarized below because it may also be interesting for the stability of dried bacterial starter cultures. A product can exist in different states depending on composition, temperature and water concentration. To simplify a rather complex matter: in a non-crystalline solid material, one can distinguish a rubbery and a glassy state. The glassy state is situated below the glass transition curve, a curve which connects the temperature-

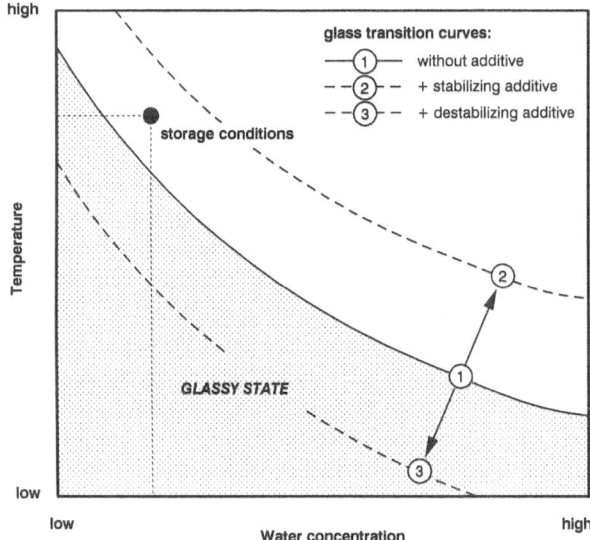

Fig. 3. A simplified state diagram which indicates the increasing or decreasing glass transition temperature after addition of stabilizing or destabilizing additives, respectively. Storage conditions are fixed by a certain product water concentration (e.g. 0.05 kg kg^{-1}) and storage temperature (e.g. 20 °C)

moisture concentration points with equal viscosity (Fig. 3). The diffusion coefficients in the glassy state are extremely low and therefore, diffusion controlled reactions become undetectable. Water in the glassy migrates also extremely slow and seems to be 'bound' or 'unfreezable' [22]. It is these low diffusion coefficients and reaction rates that will stabilize the product.

Storage at a temperature at which the product is near to or in the glassy state will thus increase the stability. To reach the glassy state, low temperatures or low water concentrations are usually needed. A much more attractive approach is to add compounds which change the position of the glass transition curve to higher temperatures, thus stabilizing the product at given storage conditions (see Fig. 3). Low DE-SHP's (low dextrose equivalent starch hydrolysis products; e.g. maltodextrines; [101]) seem to be particularly effective, but other 'sugar-like' materials are also mentioned [23, 102]. Radical reactions are not diffusion controlled and will thus not be reduced in the glassy state, however, the diffusion of oxygen into the product will be slow and this will decrease the production rate of radicals.

It must be noted that additives which stabilize the membrane (see above) can decrease the glass transition temperature. Such additives would increase survival after drying but can decrease the stability after storage. It is therefore possible, that different types of additives must be used to avoid dehydration inactivation and to increase shelf life. It seems that trehalose meets both conditions [26, 27, 104].

4 Conclusions

Damage to the cell membrane seems to be the most important dehydration inactivation mechanism but strict conclusions cannot be made. As with thermal inactivation, a combination of damage to critical and less critical sites will result in irreversible inactivation. It is likely, that the dehydration rate is not important as such, but can be important in case of harmful external conditions. Finally, it can be concluded that bacteria are most heat and dehydration resistant in the stationary growth phase. Protective additives can be applied to increase the dehydration resistance. The influence of other factors remains unclear.

An important task that remains is to investigate the mechanisms by which different factors influence survival after drying. Interdisciplinary research is needed to answer such fundamental questions. With the knowledge obtained, thermal and dehydration inactivation can then be minimized further. It is also important to obtain cultures which are stable during storage and to understand how this stability is influenced.

Acknowledgement: We would like to thank Dr. H. H. Beeftink, Mr. G. Meerdink, Dr. M. R. Smith, Dr. E. A. Tudor and Mr. M. H. Zwietering for their helpful critical comments in the preparation of this manuscript. The financial support of Gist-brocades NV, Delft, The Netherlands, is gratefully acknowledged.

5 References

1. Allwood MC, Russel AD (1970) Adv Appl Microbiol 12: 89
2. Corry JEL (1973) Progr Ind Microbiol 12: 73
3. Tomlins RI, Ordal ZJ (1976) In: Skinner FA, Hugo WB (eds) Inhibition and inactivation of vegetative microbes. Academic, London, p 153
4. Gould GW (1989) In: Gould GW (ed) Mechanisms of action of food preservation procedures. Elsevier, Essex, UK, p 11
5. Lievense LC, Verbeek MAM, Meerdink G, Van 't Riet K (1990) Bioseparation 1: 161
6. Lievense LC, Van 't Riet K (1992) In: Fiechter A (ed) Advances in Biochemical Engineering/Biotechnology, vol. 50, Springer, Berlin Heidelberg New York, p 45
7. Wagman J (1960) J Bacteriology 80: 558
8. Heckly RJ (1961) In: WW Umbreit (ed) Advances in applied microbiology. Academic Press, NY, p 1
9. Heckly RJ (1985) Develop Industrial Microbiol 26: 379
10. Webb SJ (1969) In: Nei T (ed) Freezing and drying of microorganisms. Univ Tokyo Press, Tokyo, p 153
11. Ashwood-Smith MJ (1980) In: Ashwood-Smith MJ (ed) Low temperature in medical biology. Pitman Medical, Tunbridge Wells, p 219
12. Beker MJ, Blumbergs JE, Ventina EJ, Rapoport AI (1984) Eur J Appl Microbiol Biotechnol 19: 347
13. Beker MJ, Rapoport AI (1987) In: Fiechter A (ed) Advances in Biochemical Engineering/Biotechnology, vol 32, Springer, Berlin Heidelberg New York, p 127
14. Scott WJ (1958) J Gen Microbiol 19: 624
15. Morichi T (1974) Jpn Agr Res Quart 8: 171
16. Mitic S (1976) Cryobiology 13: 214

17. Franks F (1985) In: Simatos D, Multon JL (eds) Properties of water in foods. Martinus Nijhoff, Dordrecht, The Netherlands, p 1
18. LeMaguer M (1987) In: Rockland LB, Beuchat LR (eds) Water activity: Theory and applications to food. Marcel Dekker, NY, p 1
19. Nei T (1973) Cryobiology 10: 403
20. Bousfield IJ, MacKenzie AR (1976) In: Skinner FA, Hugo WB (eds) Inhibition and inactivation of vegetative microbes. Academic, London, p 329
21. Josic D (1982) Lebensm-Wiss u Techn 15: 5
22. Franks F (1986) Cryo-Letters 7: 207
23. Franks F (1990) Cryo-Letters 11: 93
24. Brennan M, Wanismail B, Johnson MC, Ray B (1986) J Food Prot 49: 47
25. Morichi T, Irie R, Yano N, Kembo H (1967) Agr Biol Chem 31: 137
26. Crowe JH, Crowe LM, Carpenter JF, Aurell Wistrom C (1987) Biochem J 242: 1
27. Crowe JH, Carpenter JF, Crowe LM, Anchordoguy TJ (1990) Cryobiology 27: 219
28. Genis RB (1989) Biomembranes. Molecular structure and function. Springer, Berlin Heidelberg New York, p 40
29. Crowe, JH, Spargo BJ, Crowe LM (1987) Proc Natl Acad Sci (Biophysics) 84: 1537
30. Souzu H (1989) Biochim Biophys Acta 978: 105
31. Souzu H, Sato M, Kojima T (1989) Biochim Biophys Acta 978: 112
32. Bullock K, Lightbrown JW (1947) QJ Pharm Pharmacol 20: 312
33. Sobczak E (1981) Branntweinwirtschaft, July 1981: 218
34. Rogers LA (1914) J Infect Diseases 14: 100
35. Foster EM (1962) J Dairy Sci 45: 1290
36. Yankov YA, Brankova R (1979) Proc European Meat Research Workers 25 (Vol 2): 320
37. Sapp CW, Hedrick TI (1960) Q Bull Agr Exp Stat Michigan 43: 96
38. Clement P, Rossi J (1983) Preparation of dried baker's yeat. USA Patent 4.370.420. Societe Industrielle LeSaffre, Paris, France
39. Divies C, Lenzi P, Beaujeu J, Herault F (1990) Procede de preparation de micro-organismes inclus dans des gels sensiblement déshydrates, gels obtenus et leur utilisation pour la préparation de boissons fermentées. French Patent 2 633 937. Champagne Moët & Chandon, France
40. Krallish IL, Damberga BE, Beker MJ (1986) Appl Microbiol Biotechnol 23: 482
41. Krallish IL, Damberga BE, Beker MJ (1989) Appl Microbiol Biotechnol 31: 194
42. Zikmanis PB, Kruce RV, Mackare IE, Auzina LP, Beker MJ (1989) Appl Microbiol Biotechnol 31: 191
43. Gadd GM, Chalmers K, Reed RH (1987) FEMS Microbiol Lett 48: 249
44. Zikmanis PB, Auzina LP, Auzane SI, Beker MJ (1982) Appl Microbiol Biotechnol 15: 100
45. Zikmanis PB, Auzane SI, Kruce RV, Auzina LP, Beker MJ (1983) Appl Microbiol Biotechnol 18: 298
46. Zikmanis PB, Auzane SI, Auzina LP, Margevicha, Beker MJ (1985) Appl Microbiol Biotechnol 22: 265
47. Wright CT, Klaenhammer TR (1983) J Food Science 48: 773
48. Orndorff GR, MacKenzie AP (1973) Cryobiology 10: 475
49. Kilara A, Shahani KM, Das NK (1976) Cultured Dairy Products J, May 1976: 8
50. Valdez GF, De Giori GS, De Ruiz Holgodo AA, Oliver G (1983) Appl Env Microbiol 45: 302
51. Valdez GF, De Giori GS, De Riuz Holgodo AA, Oliver G (1983) Cryobiology 20: 560
52. Valdez GF, De Giori GS, De Riuz Holgodo AA, Oliver G (1985) Appl Env Microbiol 49: 413
53. Splittstoesser DF, Foster EM (1957) Appl Microbiol 5: 333
54. Webb SJ (1961) Can J Microbiol 7: 621
55. Crowe LM, Crowe JH (1988) Biochim Biophys Acta 946: 193
56. Lee CWB, Das Gupta SK, Mattai J, Shipley GG, Abdel-Mageed OH, Makriyannis A, Griffin RG (1989) Biochem 28: 5000

57. Langejan A (1980) Active dried baker's yeast. USA Patent 4.217.420. Gist-brocades, Delft, The Netherlands
58. Pomper S, Cole G, Scheinbach S (1988) Rehydratable instant active dried yeast. USA Patent 4.764.472. Nabisco Brands Inc, Parsippany, NJ, USA
59. Porubcan RS, Sellars RL (1975) Stabilized dry cultures of lactic acid producing bacteria. USA Patent 3.897.307. Chr Hansen's Lab Inc, Milwaukee, Wis, USA
60. Porubcan RS, Sellars RL (1979) In: Peppler HJ, Perlman D (eds) Microbial technology. Acacemic, London, p 59
61. De Paz M, Chavarri FJ, Nufiez M (1988) Biotechnol Techniques 2: 165
62. Bozoglu TF, Özilgen M, Bakir U (1987) Enzyme Microb Technol 9: 531
63. Schweigart F (1971) Lebensm-Wiss u Techn 4: 20
64. Freeman RR, Tschernitz JL, Marshall WR (1964) Biotechn Bioeng 6: 473
65. Masters K (1985) Spray drying handbook. George Godwin, London, p 630
66. Labuza TP, Le Roux JP, Fan TS, Tannenbaum SR (1970) Biotechn Bioeng 12: 135
67. Comings EW, Higa H, Myers JE, Koffler H, McLain HA (1977) Ind Eng Chem Fundam 16: 12
68. Espina F, Packard VS (1979) J Food Prot 42: 149
69. Marino C, Curto M, Bruno R, Rinaudo MT (1989) Int J Biochem 21: 1369
70. Zimmermann K (1987) Einflussparameter und mathematische modellierung von der schonenden trocknung von starterkulturen. Fortschr-Ber VDI (Reihe 14, no 36), VDI-Verlag, Düsseldorf, West Germany
71. Kuts PS, Tutova EG (1983) Drying Techn 2: 171
72. Lievense LC, Verbeek MAM, Taekema T, Meerdink G, Van 't Riet K (1992) Chem Eng Sci. 47: 82
73. Valdez GF, De Giori GS, De Riuz Holgodo AA, Oliver G (1985) Cryobiology 22: 574
74. Strange RE, Cox CS (1976) In: Gray TRG, Postgate JR (eds) The survival of vegetative microbes. Cambridge Univ Press, Cambridge, UK, p 111
75. Ray B, Jezeski JJ, Busta FF (1971) Appl Microbiol 22: 401
76. Hambleton P (1971) J Gen Microbiol 69: 81
77. Valdez GF, De Giori GS, De Riuz Holgodo AA, Oliver G (1985) Milchwiss 40: 518
78. Valdez GF, De Giori GS, De Riuz Holgodo AA, Oliver G (1986) Milchwiss 41: 286
79. Valdez GF, De Giori GS, De Riuz Holgodo AA, Oliver G (1985) Milchwiss 40: 147
80. Mahmoud MID, El Gammal SA, Hussein AA (1982) Zbl Mikrobiol 137: 233
81. Alaeddinoglu G, Güven A, Özilgen M (1989) Enzyme. Microb Technol 11: 765
82. Lal M, Tiwari MP, Sinha RN, Ranganathan B (1976) J Food Sci Techn India 13: 266
83. Anil Kumar PA, Gandhi DN (1982) Asian J Dairy Res 1: 283
84. Liepe HU, Junker M (1982) Arch Lebensmittelhygiene 32: 141
85. Tsetkov TS, Brankova R (1983) Cryobiology 20: 318
86. Clementi F, Rossi J (1984) Am J Enol Vitic 35: 181
87. Stadhouders J, Jansen LA, Hup G (1969) Neth Milk Dairy J 23: 182
88. Lehmann D (1984) ZFL 1984, No 2: 113
89. Hill FF (1987) Alimenta 1987 No 3: 73
90. Prajapati JB, Shah RK, Dave JM (1987) Austr J Dairy Technol March/June 1987: 17
91. Jung G (1988) Inocula of low water activity with improved resistance to temperature and rehydration and preperation thereof. USA Patent 4.755.468. Rhone-Poulenc SA, Paris, France
92. Heckly RJ (1978) In: Crowe JH, Glegg JJ (eds) Dry Biological Systems. Academic, NY, p 257
93. Sinha RN, Dudani AT, Ranganathan B (1974) J Food Science 39: 641
94. Korobkina GS, Brents MY, Kalinina NN, Vorob'eva VM, Shamanova GP (1982) Dairy Sci Abs 44: 396
95. Hill FF (1986) In: Chmiel H, Hammes WP, Bailey JE (eds) Biochemical Engineering: a challenge for interdisciplinary cooperation. Int Congress Stuttgart Sept 1986. VCH Publishers, NY, p 199
96. Kim HS, Kamara BJ, Good IC, Enders GL (1988) J Industrial Microbiol 3: 253
97. Shahani KM, Kilara A (1974) J Dairy Sci 58: 579

98. Nachmoush MR, Girgis ES, Guiguis AH, Fahmi AH (1978) Egyptian J Dairy Sci 6: 39
99. Nikolova N (1978) Proc 20th Int Dairy Congress, Paris, France: 584
100. Gehrman SH, Porubcan RS (1985) Stabilized liquid bacterial suspensions for oral administration to animals. USA Patent 4.518.696. Chr Hansen's Lab Inc, Milwaukee, Wis, USA
101. Levine H, Slade L (1989) In: Hardman TM (ed) Water and food quality. Elsevier, London, p 71
102. Franks F (1989) Process Biochem 24, No 1 (Suppl ProBioTech): R3
103. Pearce DA, Rose AH, Wright IP (1989) Yeast 5: s453 (Special issue: Proc 7th Int Symp on Yeast)
104. Roser B (1991) Trends in Food Sci Technol 2: 166

Noninferior Periodic Operation of Bioreactor Systems

S. Hasegawa[1] and K. Shimizu[2]
[1] Biotechnology Research Laboratory, Tosoh Corporation, 1-7-7 Akasaka, Minato-ku, Tokyo 107, Japan
[2] Department of Biochemical Engineering and Science, Kyushu Institute of Technology, Iizuka, Fukuoka 820, Japan

In the optimal design and operation of bioreactor systems, a single objective function has been extensively employed in the past.

Here, we formulate the optimization problem as a multi-objective programming problem which determines the noninferior solutions with respect to performance indexes. This approach has proved itself to be quite useful in getting insight into the well-balanced performance evaluation for a variety of bioreactor systems. Furthermore, we consider the optimization problem in the framework of generalized periodic operation since it contains various operations as its special cases.

List of Abbreviations and Symbols

b_i (i = 1 ~ 4)	temperature dependent parameters
E_{Ji}	J_i-extreme point
J_i	components of an objective function
Ki	inhibition constant $[\text{kg m}^{-3}]$
Ks	saturation constant $[\text{kg m}^{-3}]$
m	maintenance coefficient
$p(p_m)$	(maximum) concentration of the metabolite $[\text{kg m}^{-3}]$
$q(q_F)$	(feed) flow rate $[\text{m}^{-3}\,\text{h}^{-1}]$
RB	repeated batch operation

Advances in Biochemical Engineering
Biotechnology, Vol. 51
Managing Editor: A. Fiechter
© Springer-Verlag Berlin Heidelberg 1994

RFB	repeated fed-batch operation
$s(s_F)$	(feed) substrate concentration $[kg\ m^{-3}]$
SS	steady-state operation or continuous operation
t	time $[h]$
T_L	cultivation period counted after cell concentration reached 95% of maximum cell concentration
$v(v_m)$	(maximum) working volume $[m^3]$
x	cell concentration $[kg\ m^{-3}]$
X	$\equiv vx$
Y, Y*	yield coefficients
z	$\equiv p/(s_F - s)$
α	constant in Eq. (11)
β	constant in Eq. (11)
δ	Dirac's delta function
ε	$\equiv V*/V$
η	volume fraction of product draw-off
λ	constant for the evaluation of growth-lag period
θ	cultivation temperature $[°C]$
$\mu\ (\mu_m)$	(maximum) specific growth rate $[h^{-1}]$
ξ	volume fraction of the initial feed of substrate
π	specific production rate of metabolite $[h^{-1}]$
ϕ	distribution (partition) coefficient
τ	period $[h]$

1 Introduction

Recent developments in the biotechnology of microbial processes, recombinant DNA, hybridoma etc. have had a strong impact on the bioindustry, and there is an ever-increasing demand for further development for food, energy sources, medicine, environmental treatment etc. using microorganisms, plants, and animals.

The biotechnology industry is evolving rapidly based on the developments stated above. It seems that it has already entered a new stage of growth and is about to leave behind the early stage of skepticism. The many biotechnology-based products such as pharmaceutical and health-care products, agricultural products, and chemicals have already been commercialized and attention is being focused on the transition from laboratory to marketplace [1]. Billions of dollars are being invested annually by many companies looking for capital appreciation and seeking business opportunities.

Despite the steady progress in laboratory-scale research, however, there remain many problems associated with the scale-up of bioprocesses. Since most biochemical processes create very dilute and impure products, there is a great need to increase volumetric productivity and to increase the product concentration etc. As a result, the scale-up of bioprocesses requires a large investment in developing an efficient processing technology. In this regard, the significant work is needed to optimize the design and operation of bioreactors to make production more efficient and more economical.

Some of the basic ideas for the better design and operation of bioreactors have been described so far by many researchers [2–9]. In the investigations on the optimal design and operation, a single objective function has been extensively used among productivity, product concentration, substrate conversion, cost of utilities etc. However, from the standpoint of the optimization for the total process including up-stream and down-stream processes, it is necessary to take the above performance indexes into account simultaneously.

With the above-stated background, it may be quite useful to analyze the optimal operation mode using a vector-valued objective function taking into account several performance indexes simultaneously. The optimization problem can be formulated as a multiobjective programming problem which determines the noninferior solutions with respect to performance indexes. This approach has proved useful in getting insight into well-balanced performance evaluation.

It should be noted that the performance of a bioreactor is highly dependent on the choice of the operational mode. The determination of the most suitable operational mode is, however, not easy, since different optimization techniques are required for different operational modes in general, and the optimal operation mode can be determined only after the optimization was made for each candidate mode. As a result, a vast amount of computation is required to find the optimal operation mode.

To overcome this problem, we considered the optimization problem in the framework of generalized periodic operation, since the periodic operation can be regarded as containing various possible types of operational modes as its special cases.

In the following, we review the work done for the performance evaluation of bioreactor systems using a vector-valued objective function in the framework of generalized periodic operation.

2 Biomass Producing Systems

2.1 Problem Formulation

Consider first the simple situation where the dynamic behavior of the bioreactor is represented by the following equations for the cell mass production:

$$d(vx)/dt = -qx + \mu(s)vx \tag{1a}$$

$$d(vs)/dt = q_F s_F - qs - \mu(s)vx/Y(s) \tag{1b}$$

$$dv/dt = q_F - q \tag{1c}$$

where x, s, v are the cell concentration, substrate concentration, and medium volume, respectively. q_F and q are the feed rate and product draw-off rate, respectively. s_F is the feed substrate concentration. The specific growth rate μ and the yield coefficient Y may typically be given as functions of s as

$$\mu(s) = \mu_m s/(Ks + s + s^2/Ki) \tag{2a}$$

$$1/Y(s) = 1/Y^* + m/\mu(s) \tag{2b}$$

Consider a periodic operation of the reactor in which $x, s,$ and v are in a periodic state with period τ subject to a periodic change in q_F and q. s_F is assumed to be constant. Note that the steady-state operation or continuous operation can be regarded as the special case of the periodic operation since any constant-valued function is periodic with respect to an arbitrary value of a period.

Consider the cell productivity and substrate conversion as the components of a vector-valued objective function, where those are expressed by

$$J_1 = \int_0^\tau qx \, dt/v_m \tau \tag{3a}$$

and

$$J_2 = \int_0^\tau qx \, dt/\int_0^\tau q_F s_F \, dt \tag{3b}$$

where v_m is the maximum working volume of the reactor.

Let (J_1^*, J_2^*) be a feasible value of the vector-valued objective function (J_1, J_2). Then the solution (J_1^*, J_2^*) is said to be noninferior if, and only if, there exists no feasible (J_1, J_2) satisfying either $J_1 \geq J_1^*$ and $J_2 > J_2^*$, or $J_1 > J_1^*$ and $J_2 \geq J_2^*$. Note that the value of J_1 is maximized at one extreme point E_{J1} of the noninferior set and the value of J_2 is maximized at the other extreme point E_{J2}.

Assume for simplicity that the product is drawn off only at the end of each cycle for the general periodic operation. If the fraction η of the reactor content is drawn off as product, it follows that

$$q(t) = \eta v_m \delta(t - \tau) \tag{4}$$

where δ is Dirac's delta function.

Then it can be shown that

$$J_1 = \eta x(\tau)/\tau \tag{5a}$$

$$J_2 = x(\tau)/s_F \tag{5b}$$

2.2 Noninferior Set

The noninferior set can be determined by use of the constraint method [10] in which one-objective programming problem is solved repeatedly for all the feasible values of J_2^*. If the value of J_2 is specified, the value of $x(\tau)$ is fixed accordingly. If in addition the value of η is specified, the optimization problem is reduced to a minimal-time problem in which the value of τ is minimized for specified values of η and $x(\tau)$. The same minimal-time problem was studied by Weigand [11] using Pontryagin's maximum principle and the generalized Legendre-Clebsch condition. However, the more expedient condition to this problem is to invoke the optimization technique of Miele [12], which was successfully applied by Yamane et al. [13] to the optimal start-up problem of chemostat culture.

It follows that

$$\tau = \int_{X(0)}^{X(\tau)} \frac{dX}{\mu(s)X} \tag{6}$$

where $X \equiv vx$.

Let Γ be a simple closed curve in the Xs-plane and Σ the region surrounded by Γ. Then it follows from Green's theorem that

$$\oint_\Gamma \frac{dX}{\mu(s)X} = \iint_\Sigma \frac{\mu'(s)}{\mu^2(s)X} \, ds \, dX \tag{7}$$

where the path of the line integral in the left-hand side must be taken counterclockwise.

Any feasible trajectories connecting the points I and F must be contained in the region IAFB as indicated in Fig. 1, where the branches AF and IB corres-

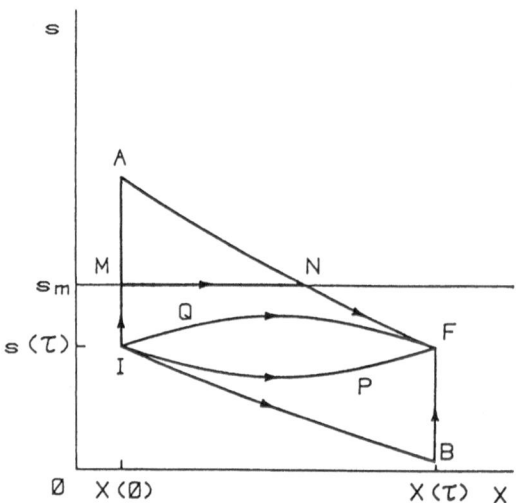

Fig. 1. Minimal-time trajectory in the Xs-plane ($s(\tau) < s_m$) (From Ref. 14)

pond to $q_F = 0$ while the branches IA and BF correspond to $q_F \rightarrow \infty$. Let IPF and IQF be two trajectories contained in the region IAFB. If the curve IPF is located in the lower side of IQF, then it follows from Eq. (7) that

$$\tau_{IPF} - \tau_{IQF} = \oint_{IPFQI} \frac{dX}{\mu(s)X} = \int\int_{IPFQI} \omega(X, s)\, dX ds \qquad (8)$$

where ω is the integrand in the RHS of Eq. (7). Note that the sign of ω is determined by that of $\mu'(s)$. Let s_m be the value of s when $\mu'(s) = 0$ holds. Then it can be said that $\tau_{IPF} > \tau_{IQF}$ if IPFQI is contained in the region of $0 \leqq s \leqq s_m$ while $\tau_{IPF} < \tau_{IQF}$ if IPFQI is contained in the region of $s \geqq s_m$. Thus, it is easy to determine the minimal-time trajectory of interest. Yamane *et al.* [13] adopted the Xμ-plane in place of the Xs-plane. Since, however, two values of s correspond in general to any points in the former plane, the latter seems more advantageous than the former for applying Miele's technique.

   If $s(\tau) < s_m$ as shown in Fig. 1, the minimal time trajectory is IMNF, which reduces to IAF when the point A falls in the region of $0 \leqq s \leqq s_m$. The branches IM, MN, and NF of the trajectory IMNF represent an instantaneous fill of the substrate, a fed-batch operation, and a batch operation, respectively. The trajectory IAF consists of an instantaneous fill branch IA and a batch operation branch AF. The periodic operations corresponding to the trajectories IMNF and IAF are usually called repeated fed-batch and repeated batch operation, respectively.

   It can be shown [14] that the noninferior set for the general periodic operation consists of the branches of the repeated batch and the repeated fed-batch operation, and the latter branch disappears in the case without substrate inhibition. It can be also shown that the high-productivity portion of

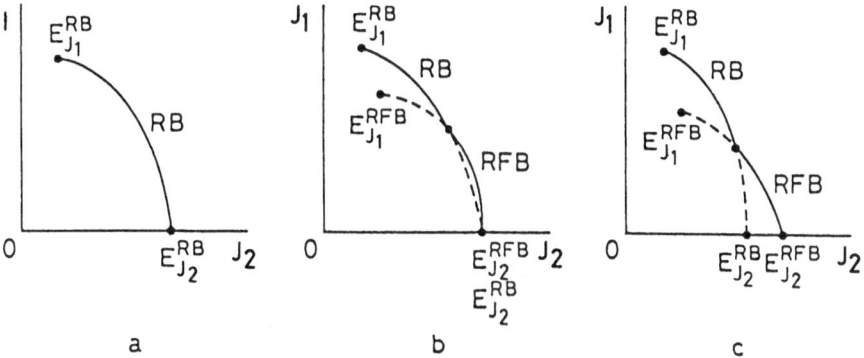

Fig. 2a–c. Typical possible structures of noninferior set. (a) Case of $Ki \to \infty$. (b) Case of $m = 0$. (c) Case of $m > 0$ (From Ref. 14)

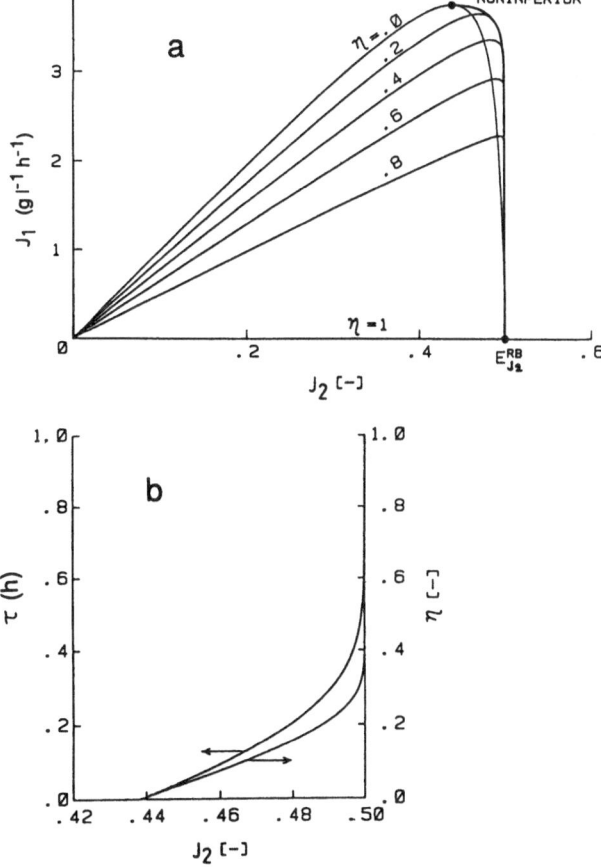

Fig. 3a. Noninferior set and J_1 vs J_2 curves for various specified value of η. (b) The η and τ vs J_2 curves for noninferior solutions (From Ref. 14)

the noninferior set is occupied by the repeated batch branch, and that the J_1-extreme point is the point E_{J1}^{RB}, J_1 extreme point for repeated batch operation, which agrees with the point E_{J1}^{SS}, J_1 extreme point for steady-state operation.

In Fig. 2, typical possible structures of the noninferior set are illustrated. It can be seen for the substrate-inhibition kinetics that the high conversion portion of the noninferior set is occupied by the repeated fed-batch branch.

It is seen in Fig. 3(a) that the value of J_2 can be increased by use of the repeated batch operation at the cost of a small decrease in the value of J_1 relative to the steady-state operation. The values of η and τ increase with increase in J_2 as indicated in Fig. 3(b).

Once the noninferior set was determined, one could choose the preferable operation mode by the trade-off between the cell productivity and the substrate conversion. It has been well known prior to this work that the cell productivity is maximized by the steady-state operation and that the repeated batch or the repeated fed-batch operation is superior to the steady-state operation if higher substrate conversion is desired. In this work, it was made clear by investigating the structure of the noninferior set that much increase in the conversion could in general be obtained at the cost of a relatively small decrease in the cell productivity, and that the repeated batch operation should be adopted in place of the repeated fed-batch operation if higher cell productivity was desired in the case of substrate-inhibited growth kinetics [14].

3 Metabolite Producing Systems

3.1 Problem Formulation

In 1959, Gaden [15] classified the metabolites of the microbial processes into three types from the kinetic view point,

Type I: growth-associated products arising directly from the energy metabolism of carbohydrates supplied,

Type II: indirect products of carbohydrate metabolism, and

Type III: products apparently unrelated to carbohydrate oxidation.

The periodic operation of the bioreactor producing the metabolites of type III will be dealt with in the next section. Consider here the problem where the produced metabolite belongs to Type I or II. Typical examples of the metabolites of Type I are ethanol, lactic acid etc., while those of Type II are citric acid, salicylic acid etc.

Consider a well-mixed bioreactor in which some metabolites are produced by cells. The dynamic behavior may be described by

$$d(vx)/dt = -qx + \mu(s, p)\,vx \qquad (9a)$$

$$d(vs)/dt = q_F s_F - qs - \mu(s, p)vx/Y \tag{9b}$$

$$d(vp)/dt = - qp + \pi(s, p)vx \tag{9c}$$

$$dv/dt = q_F - q \tag{9d}$$

where p represents the concentration of the desired metabolite.

The specific growth rate may be expressed in many cases as

$$\mu(s, p) = \{\mu_m s/(Ks + s + s^2/Ki)\}(1 - p/p_m) \tag{10}$$

and the specific metabolite production rate may be given by the well-known Luedeking-Piret model [16] such as

$$\pi(s, p) = \alpha\mu(s, p) + \beta \tag{11}$$

Here, β is nonnegative, while α can be negative if β is positive. The case where α is negative was reported for the glutamic acid cultivation by *Brevibacterium* sp. [17]. The special case where $\alpha > 0$ and $\beta = 0$ corresponds to the case of the metabolite production of Type I.

If we can assume s_F and Y to be constant, the stoichiometric relation

$$x = Y(s_F - s) \tag{12}$$

holds for the entire course of cultivation if it does at the initial time. By using this relation, the state variable x can be eliminated from Eq. (9).

Here, we formulate the optimization problem as a three-objective programming one with respect to the productivity of the desired metabolite, J_1, its concentration, J_2, and the substrate conversion, J_3.

Note that the point (J_1^*, J_2^*, J_3^*) is said to be noninferior if, and only if, there exists no feasible point (J_1, J_2, J_3) satisfying $J_i \geqq J_i^*$, $i = 1, 2, 3$ except the point itself. In general, each objective function will be maximized at different points within the noninferior set. As before, the extreme points are called J_i-extreme points $(i = 1, 2, 3)$ and are designated by the symbols E_{J1}, E_{J2}, and E_{J3}, respectively.

3.2 Noninferior Set

Consider first the steady-state operation. It can be shown [18] that the feasible set and the noninferior set in $J_1 J_2 J_3$-space may indicate various patterns for the values of kinetic parameters in a complex way. Typical examples of the feasible set and the corresponding noninferior set are depicted in Fig. 4 for $p_m < \infty$. The solid lines indicate the feasible set and the thick lines indicate the corresponding noninferior set.

Consider next the case of general periodic operation. We assume that the product is drawn off quickly only at the end of the period. The volume of the reactor content just before the draw-off of the product should be equal to v_m by which the productivity is maximized. Let η be the ratio of the draw-off volume

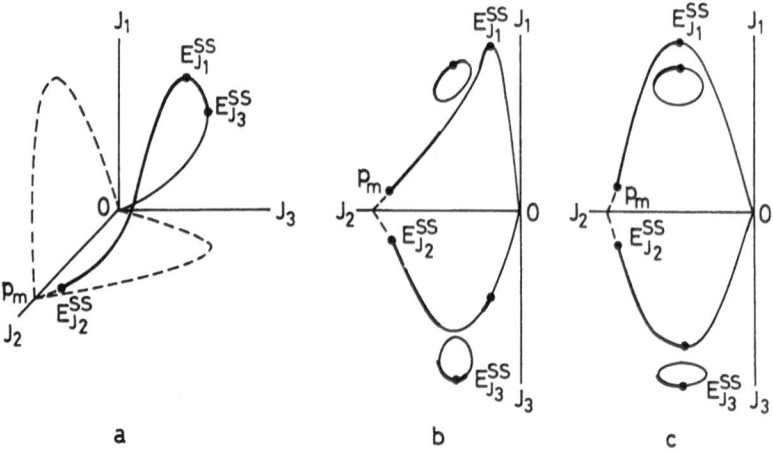

Fig. 4. Typical possible patterns of the feasible set and the corresponding noninferior set for the steady-state operation ($p_m < \infty$) (From Ref. 18)

to v_m, and assume that the substrate of the volume ξv_m ($0 < \xi \leq \eta$) is quickly fed at the initial time. Then each objective function can be expressed as

$$J_1 = \eta p(\tau)/\tau \tag{13a}$$

$$J_2 = p(\tau) \tag{13b}$$

$$J_3 = \{s_F - s(\tau)\}/s_F \tag{13c}$$

The noninferior set can be determined by extracting the noninferior portion from the manifold in the $J_1 J_2 J_3$-space which is obtained by repeating the procedure of maximizing J_1 for a feasible set of the values J_2 and J_3, or equivalently the values of $p(\tau)$ and $s(\tau)$. Then the maximization of J_1 reduces to that of η/τ. The latter maximization can be achieved by maximizing with respect to η the solution of the minimal-time problem in which τ is minimized for a specified value of η. This minimal time problem can be solved by Miele's method as mentioned previously.

In the application of Miele's method to the minimal-time problem, it is convenient to introduce a new state variable z defined by

$$z \equiv p/(s_F - s) \tag{14}$$

When $q_F(t)$ and η are given, $v(t)$ can be obtained as the solution to Eq. 9(d). Thus, the three-dimensional system for s, z, and v can be regarded as the two dimensional one with respect to z and s. It can be shown [18] that

$$\tau = \int_{z(0)}^{z(\tau)} \frac{dz}{\pi(s, z)Y - \mu(s, z)z} \tag{15}$$

Suppose that a simple closed curve Γ is composed of two trajectories APB and

AQB starting from the point A and ending at point B, and that the direction of the trajectory APB is that of going around Γ counterclockwise. Let Σ be the region surrounded by Γ. Then it follows from Green's theorem that

$$\int_{APB} dt - \int_{AQB} dt = \oint_\Gamma \frac{dz}{\pi Y - \mu z} = \iint_\Sigma \frac{\omega(z, s)}{(\pi Y - \mu z)^2} \tag{16}$$

where

$$\omega(z, s) \equiv (\alpha Y - z)\mu_s \quad \text{and} \quad \mu_s \equiv \partial\mu/\partial s$$

We assume here that $\pi Y - \mu z$ does not identically vanish for any finite time interval.

The trajectory which corresponds to a periodic state must be a closed curve in the zs-plane because of the boundary condition. The directions of the state change in the zs-plane are determined by ds/dt and dz/dt. It can be shown [18] that the sign of ds/dt depending on the value of q_F is positive for $q_F > \mu v$, while negative for $q_F < \mu v$. On the other hand, the sign of dz/dt is independent of q_F. From Eqs. (9), (11), and (14), the following equation holds:

$$dz/dt = (\alpha Y - z)\mu + \beta Y \tag{17}$$

It can be seen that $dz/dt > 0$ holds in the region where $z \leq \alpha Y$ if we consider the case where $\beta > 0$. In the region where $z > \alpha Y$, the locus $dz/dt = 0$ denoted by Λ exists. Typical shapes of Λ are indicated in Fig. 5.

It can be shown [18] that the closed trajectory is counterclockwise (respectively clockwise) if dz/dt is positive (respectively negative) for the portion of the branch BI in the neighborhood of the point I (see Fig. 6). In such a situation, the upper portion of the closed trajectory IABI must be lying in the region where $dz/dt < 0$ (respectively $dz/dt > 0$). In other words, the branches IA and ABI must intersect the locus Λ. The location of I or the values of J_2 and J_3 for the possibility of the closed trajectory is restricted by this condition.

Points A and B should be so determined as to be optimal. Consider the case of the counterclockwise trajectory as shown in Fig. 6(a). Suppose that the extensions of IA and IB meet at point C. Let A' and B' be points lying on AC and BC, respectively. If $\omega < 0$ holds in the whole interior of the closed curve AA'QB'BPA, then

$$\int_{AA'B'B} dt < \int_{APB} dt$$

while if $\omega > 0$ holds in place of $\omega < 0$, then

$$\int_{AA'B'B} dt > \int_{APB} dt$$

This means that it is beneficial to move the branch AB upward in the region where $\omega < 0$ as much as possible.

If the whole interior of the closed curve IACBI is contained in the region where $\omega < 0$, then the closed curve is the optimal trajectory. However, if curve

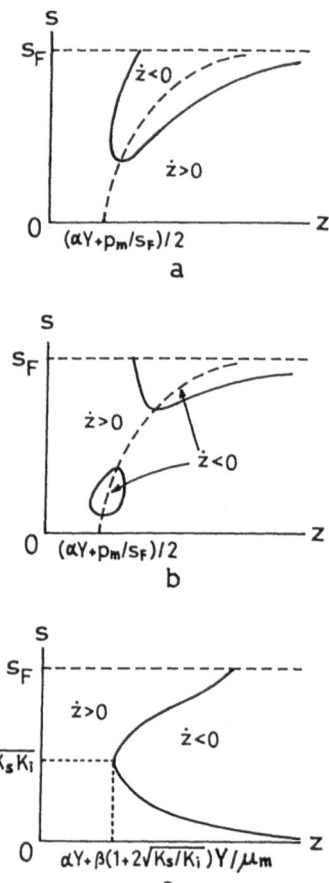

Fig. 5a–c. Possible types of locus Λ. (a) and (b) Case of $p_m < \infty$. (c) Case of $p_m \to \infty$ (From Ref. 18)

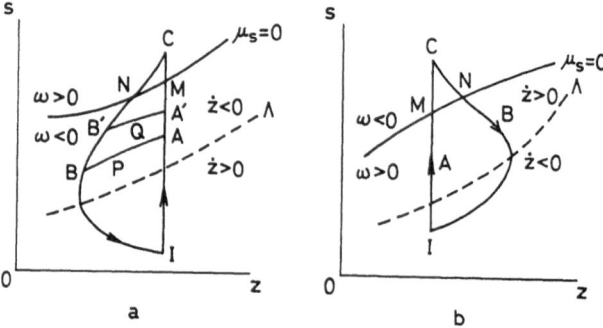

Fig. 6. The minimal-time trajectories in the zs-plane (From Ref. 18)

$\mu_s = 0$ intersects IAC and CBI at points M and N, respectively, and if $\omega < 0$ holds in the lower side of $\mu_s = 0$, the optimal trajectory is IMNI whose branch MN passes along the curve $\omega = 0$. If ω is positive for the portion of BI near point I, then it is optimal to contract the closed trajectory to point I which corresponds to the steady-state operation. Evidently, this operation is feasible only if point I is located just on the locus Λ.

For the clockwise trajectory as shown in Fig. 6(b), similar discussion can be made by reversing the conditions concerning the sign of ω.

3.3 Simulation

The simulation results are shown in Fig. 7(a). The broken line shows the upper boundary of the feasible set for the periodic operation and the corresponding noninferior set is shown by the thick line. In Fig. 7(b), the values of η, ξ, and

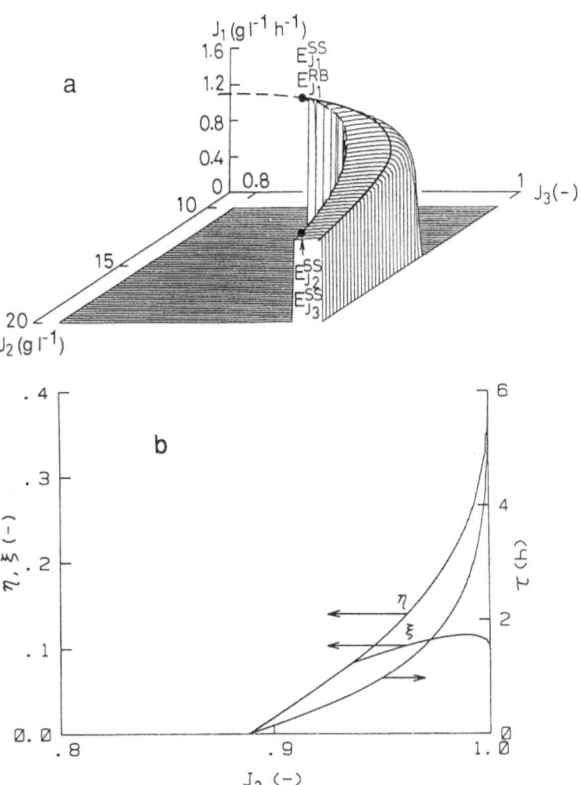

Fig. 7a. The feasible set for the steady-state operation and the noninferior set for the periodic operation. **(b)** The value of η, ξ, and τ corresponding to the noninferior set (From Ref. 18)

τ corresponding to the noninferior set are depicted. In consequence, the non-inferior operation is the repeated batch and repeated fed-batch, respectively. Anyway, the noninferior set consists of two portions corresponding to the repeated batch operation and the repeated fed-batch operation. At the J_1-extreme point E_{J1}^{RB} for the repeated batch operation, both η and τ tend to zero and the repeated batch operation becomes equivalent to the steady-state operation.

From the simulation results, it can be said that near the productivity-extreme point, the optimal operation became the repeated batch operation and at that point it became equivalent to the steady-state operation. When the higher metabolite concentration and/or the higher substrate conversion was desired, the optimal operation might become the repeated fed-batch operation. It was clarified that, in general, the concentration of metabolite and/or the substrate conversion could be increased at the cost of a relatively small decrease in the productivity of the desired metabolite by adopting an adequate periodic operation.

4 Multiple Bioreactor Systems

Production of L-glutamic acid is an example of Type III in Gaden's classification, and the production of secondary metabolites such as antibiotics and giberellin is a typical example of this case. In these cases, the product is rather expensive so that various efforts have been focused on increasing the productivity and the product concentration. The direct application of repeated batch or fed-batch cultivation to such systems, however, may not lead to an improvement in productivity. The reason is as follows: In order to obtain a higher concentration of the product, one must wait long after the time t_x, say t_r, in Fig. 8. However, if a repeated batch or fed-batch operation is conducted with an inoculation of x_r in Fig. 8, a longer lag phase might result since the inoculated cells are old. Thus the productivity cannot be increased.

We proposed using multiple bioreactors to overcome this problem [19], and showed the usefulness of this idea for the efficient production of penicillin by computer simulation using the following model proposed by Constantinides et al. for batch operation [20, 21].

$$dx(t)/dt = b_1(\theta)x(t)\{1 - x(t)/b_2(\theta)\} \tag{18a}$$

$$dp(t)/dt = b_3(\theta)x(t - 20) - b_4(\theta)p(t) \tag{18b}$$

where x and p represent cell and penicillin concentrations, respectively. θ is cultivation temperature, and b_i $(i = 1 \sim 4)$ are temperature dependent parameters.

The idea can be described by considering the repeated batch operation as follows (see Fig. 9):

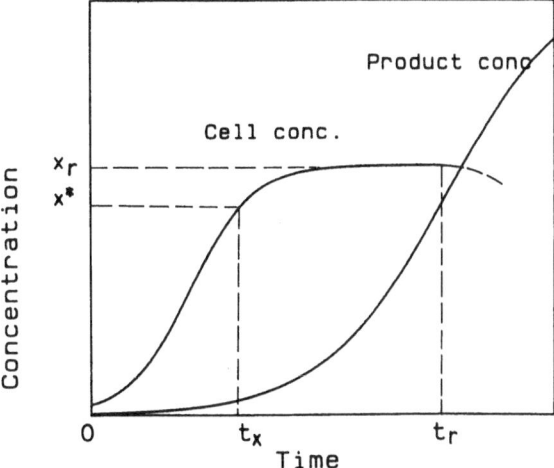

Fig. 8. Illustration of typical cell growth and product formation curves: x, cell concentration (From Ref. 19)

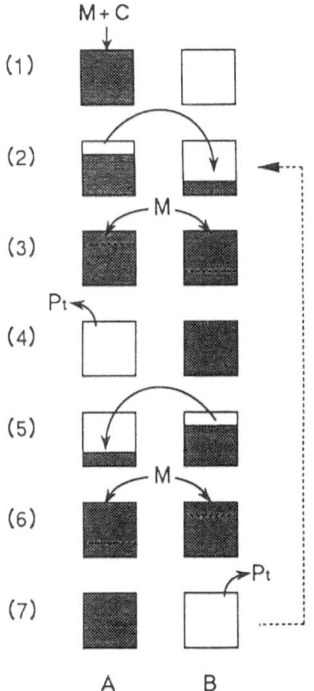

Fig. 9. Operating procedure for the repeated batch operation using two bioreactors: *M*, fresh medium; *C*, cells; *Pt*, product (From Ref. 19)

(1) Cells are inoculated into bioreactor A which is filled with nutrient, and the cultivation is started.
(2) When the cell concentration x, reaches some specified value x*, or when it has been cultivated for a duration T, then a fraction η of the culture medium is transferred into bioreactor B.
(3) Both bioreactor A and B are filled with fresh medium to the maximum working volume. Bioreactor A continues cultivation and bioreactor B starts cultivation.
(4) When the cell concentration in bioreactor B reaches x*, or it is cultivated for T, all the contents of bioreactor A are harvested.
(5) Then the fraction η of the contents of bioreactor B is transferred into bioreactor A.
(6) Fresh medium is added into both bioreactors to their maximum working volume, and cultivation is continued.
(7) When the cell concentration of bioreactor A reaches x*, or when cultivation time reaches T, all the contents of bioreactor B are harvested.

The above procedure is repeated from step (2).

By the above operation, younger cells can be used for inoculation of the repeated batch (or fed-batch) operation. In the simulation, we also assumed that the growth-lag period of cell at inoculation can be expressed as a linear function of cultivation age, T_L, counted after the cell concentration reaches 95% of the maximum cell concentration attainable in batch cultivation under the optimal temperature as λT_L.

Figure 10 shows the simulation result where biobjective programming problem was solved by maximizing the productivity with the specified penicillin concentration. The calculation was made by seeking the optimal condition, that is x*, θ, and η, for productivity for a specified harvesting penicillin concentration. Each curve shows the so-called noninferior set for each operation. The noninferior set varies with the value of λ for the repeated batch operation using one bioreactor, but is affected only slightly for the repeated batch operation using two bioreactors. It can be seen from the global noninferior set shown in the figure that the repeated batch operation using two bioreactors is the most efficient operation mode among the three if the specified penicillin concentration is high.

The idea of using multiple bioreactors can be successfully applied to many other processes. We applied this idea to improve the efficiency of vinegar production, and showed the usefulness of using multiple bioreactors by comparing the noninferior sets [22]. In vinegar production, ethanol is supplied as the main substrate, and acetic acid is produced as the main metabolic product. It has been long known that ethanol and acetic acid both inhibit the growth of *Acetobacter* sp. Concerning the growth inhibition due to ethanol, it has been shown that the cell growth is little affected if the ethanol concentration in the bioreactor is kept at around 5–30 g l^{-1}, which is not difficult to attain in practice by the conventional fed-batch mode of operation.

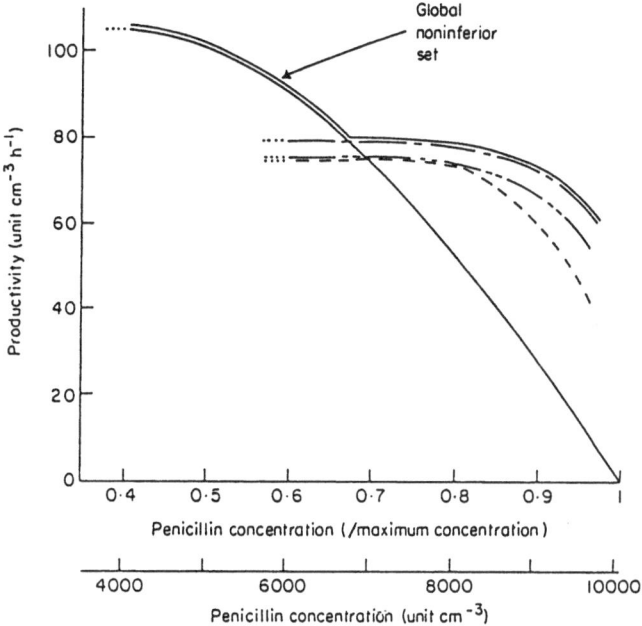

Fig. 10. Noninferior set of biobjective programming problem on penicillin cultivation: ———, steady-state operation; —·—, repeated batch operation using two bioreactors; — — —, repeated batch operation using one bioreactor ($\lambda = 0.3$); —··—, repeated batch operation using one bioreactor ($\lambda = 0.6$) (From Ref. 20)

The problem is the growth inhibition due to the accumulation of acetic acid. In the conventional fed-batch operation, the productivity has been limited low, since the growth inhibition becomes quite significant when the acetic acid concentration increases to more than about 50 g l^{-1}. It should be noted that the acetate concentration should be more than 80 g l^{-1} in practice.

Employing the repeated-batch operation may be considered for increasing the productivity. However, as has been pointed out by Park et al. [23], the number of viable cells tends to decrease abruptly as the acetic acid concentration increases to more than about 60 g l^{-1}. This means that an improvement in the productivity can hardly be expected if the culture broth is harvested when the acetic acid concentration reaches about 80–90 g l^{-1} and part of the culture medium is used as the inoculation for the next cultivation.

It should be recalled that the acetate concentration must be higher than about 80 g l^{-1}, while the number of viable cells in the medium to be used as the inoculation should be high. These contradictory requirements can be satisfied by the repeated fed-batch operation using multiple bioreactors.

5 Extractive Cultivation Systems

One of the major problems which prevent us from attaining high productivity in microbial processes is the end product inhibition of the microorganism involved due to the accumulation of toxic metabolites such as acetone, butanol, ethanol etc. in the bioreactor. Since the accumulation of toxic products slows down and finally stops the growth of microorganisms, the productivity attainable has been rather limited.

A reasonable approach to increasing productivity may be to remove the toxic products as they are formed. One promising method of removing toxic metabolites is to make use of liquid-liquid extraction [24–26]. The key to the success of this approach is to find the appropriate solvents which can effectively extract toxic products during cultivation. Significant efforts to find appropriate solvents demonstrates that this approach is becoming more and more attractive [27, 28].

With the above-stated background, we have developed a general framework for the assessment of extractive cultivation and showed a significant perform-ance improvement, in particular for acetone-butanol extractive cultivation using oleyl alcohol as an extractant [29].

The basic equations which describe the batch extractive cultivation may be expressed as

$$dx/dt = \mu(s, p_1, p_2, \ldots, p_n)x \tag{19a}$$

$$ds/dt = -\mu(s, p_1, p_2, \ldots, p_n)x/Y \tag{19b}$$

$$d(p_i V + p_i^* V^*)/dt = \mu(s, p_1, p_2, \ldots, p_n)xV/Y_i \quad (i = 1, 2, \ldots, n) \tag{19c}$$

where x and s are the cell and substrate concentrations, respectively. V is the reactor volume. In Eq. (19c), the asterisk refers to the solvent phase. p_i is the i-th metabolic product among n metabolites and Y_i denote the yield coefficients defined as

$$Y_i \equiv \Delta x/\Delta p_i \quad (i = 1, 2, \ldots, n) \tag{20}$$

and are assumed to be constant.

The specific growth rate is assumed to be of the form

$$\mu(s, p_1, p_2, \ldots, p_n) = \frac{\mu_m s}{Ks + s} \prod_{i=1}^{n} (1 - p_i/p_{im})^{q_i} \tag{21}$$

In the following, we assume that $q_i = 1$ $(i = 1, 2, \ldots, n)$ for simplicity without loss of practical significance [30, 31].

The distribution (partition) coefficient, ϕ_i, is defined as the ratio of the concentration of the i-th metabolite in the solvent phase to that in the medium (water) phase such that

$$p_i^* = \phi_i p_i \quad (i = 1, 2, \ldots, n) \tag{22}$$

It may be reasonable to evaluate the performance of extractive cultivation systems by the productivity and the concentration of a metabolite, butanol in this case. The objective functions J_1 and J_2 for repeated batch operation are then given by

$$J_1 = \frac{p_i(\tau)V + p_i^*(\tau)V^*}{(V + V^*)\tau} = \frac{(1 + \phi_i \varepsilon)p_i(\tau)}{(1 + \varepsilon)\tau} \tag{23a}$$

$$J_2 = \frac{p_i(\tau)V + p_i^*(\tau)V^*}{V + V^*} = \frac{(1 + \phi_i \varepsilon)p_i(\tau)}{1 + \varepsilon} = J_1 \tau \tag{23b}$$

where $\varepsilon \equiv V^*/V$ and p_i corresponds to butanol concentration. τ is the batch operation period. Note that the following conditions hold for the repeated batch operation as part of the medium $\eta V (0 < \eta < 1)$ and all of the solvent are drawn off at time τ and then fresh medium of η V and solvent of $V^* (= \varepsilon$ V) are added.

$$s(0) = (1 - \eta)s(\tau) + \eta \, s_F \tag{24a}$$

$$x(0) = (1 - \eta)x(\tau) \tag{24b}$$

$$p_i(0) = (1 - \eta)p_i(\tau)/(1 + \phi_i \varepsilon) \quad (i = 1, 2, \ldots, n) \tag{24c}$$

Figure 11 shows the simulation result for the batch operation ($\eta = 1$) for various values of ϕ. It clearly shows the significant performance improvement for the extractive cultivation using oleyl alcohol ($\phi = 3.7$).

Figure 12 shows the simulation result for the repeated batch operation, where the optimization was carried out by Nelder and Mead's simplex method [32] with respect to η and ε for the specified values of J_2. Here the substrate concentration was assumed to be limited to the range between 0 and 80 g l^{-1}, because substrate inhibition occurs in practice for acetone-butanol cultivation [27]. As can be seen from Fig. 12, a significant improvement is observed in comparison with the result of batch operation. On the other hand, the maximum value of J_2 was below that obtained for the batch operation. This is due to the upper limitation of the substrate concentration.

Note that in such microbial processes as acetone-butanol cultivation, more than one metabolite such as acetone, butanol, ethanol etc. is formed during cultivation each affecting the growth of the microorganism in a complex way. In the extractive cultivation using oleyl alcohol as an extractant, the primary toxic metabolite, such as butanol, is selectively extracted from the medium. As the butanol is extracted, however, another metabolite in the medium such as acetone becomes the next critical metabolite which slows down and finally stops the growth. With this in mind, it is quite reasonable to consider the more efficient extractive cultivation strategy using multiple solvents each extracting a different metabolite selectively. Since the microbial process is multicomponent in nature, the theoretical development is highly complicated.

We have made, a detailed, yet general mathematical formulation, where two types of solvent-supplying strategies were considered. One is to add multiple solvents simultaneously and the product is removed at one time. Another is to

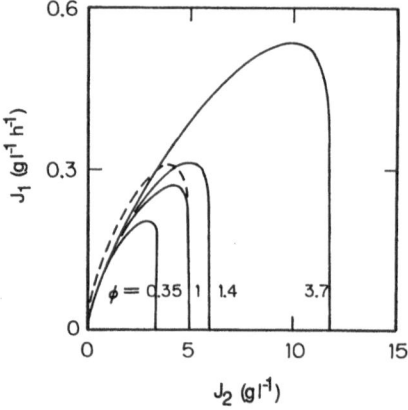

Fig. 11. Comparison of performances for batch operation ($\eta = 1$): ——, extractive cultivation; — — —, conventional cultivation without extraction (From Ref. 29)

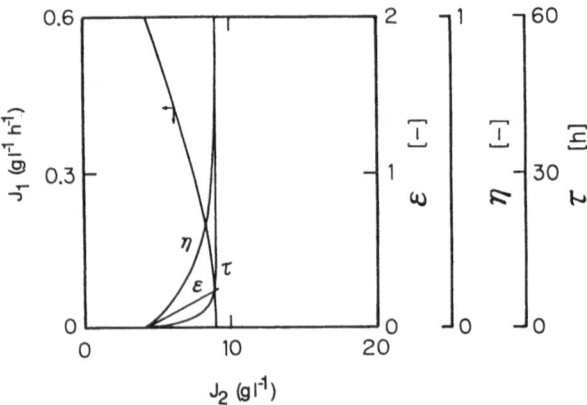

Fig. 12. Performance evaluation of the extractive repeated batch operation for a acetone-butanol process (From Ref. 29)

add them one by one consecutively. This was applied to batch, fed-batch, and repeated fed-batch operation of acetone-butanol cultivation to show the power of the approach. The results show that the significant performance improvement in terms of the productivity and product concentration can be attained when two extractants such as oleyl alcohol and benzyl benzoate were used as compared with the case of using only one solvent [33].

6 References

1. Shamel RE, Chow JJ (1987) Chem Eng Prog 10: 41
2. Aiba S, Humphrey AE, Millis NF (1973) Biochemical engineering, 2nd edn. Academic, New York
3. Atkinson B (1974) Biochemical reactors. Piton, London
4. Pirt SJ (1975) Principles of microbe and cell cultivation. Blackwell, Oxford
5. Bailey JE, Ollis DF (1977) Biochemical engineering fundamentals. McGraw-Hill, New York
6. Wang DIC, Cooney CL, Demain AL, Dunhill P, Humphrey AE, Lilly MD (1979) Fermentation and enzyme technology. Wiley, New York
7. Benefield LB, Randall CW (1980) Biological process design for wastewater treatment. Prentice-Hall, Englewood Cliffs
8. Schügerl K (1987) Bioreaction engineering, vol 1. Wiley, New York
9. Moser A (1988) Bioprocess technology: Kinetics and reactors. Springer, Berlin Heidelberg New York
10. Cohon JL (1978) Multiobjective programming and planning, Academic, New York
11. Weigand WA (1981) Biotechnol Bioeng 23: 249
12. Miele A (1962) In: Leitman G (ed) Optimization techniques, Academic, New York
13. Yamane T, Sada E, Takamatsu T (1979) Biotechnol Bioeng 21: 111
14. Matsubara M, Hasegawa S, Shimizu K (1985) Biotechnol Bioeng 27: 1214
15. Gaden Jr EL (1959) J Biochem Microbiol Technol Eng 1: 413
16. Luedeking R, Piret EL (1959) J Biochem Microbiol Technol Eng 1: 393
17. Shingu H, Terui G (1971) J Ferment Technol 49: 400
18. Hasegawa S, Matsubara M, Shimizu K (1987) Biotechnol Bioeng 30: 703
19. Shimizu K, Kobayashi T, Nagara A, Matsubara M (1985) Biotechnol Bioeng 27: 743
20. Hasegawa S, Shimizu K, Kobayashi T, Matsubara M (1985) J Chem Tech Biotechnol 35B: 33
21. Constantinides A, Spencer JL, Gaden Jr EL (1970) Biotechnol Bioeng 12: 1081
22. Ito T, Sota H, Honda H, Shimizu K, Kobayashi T (1991) Appl Microbiol Biotechnol 36: 295
23. Park YS, Toda K, Fukaya M, Okuhara H, Kawamura Y (1991) Appl Microbiol Biotechnol 35: 149
24. Wang HY, Robinson FM, Lee SS (1981) Biotechnol Bioeng Symp :555
25. Minier M, Goma G (1982) Biotechnol Bioeng 24: 1565
26. Matsumura M, Markle H (1984) Appl Microbiol Biotechnol 20: 371
27. Taya M, Ishii S, Kobayashi T (1985) J Ferment Technol 63: 181
28. Ishii S, Taya M, Kobayashi T (1985) J Chem Eng Japan 18: 125
29. Honda H, Mano T, Taya M, Shimizu K, Matsubara M, Kobayashi T (1987) Chem Eng Sci 42: 493
30. Holzberg I, Finn RK, Steinkraus KH (1967) Biotechnol Bioeng 9: 413
31. Ghose TK, Tyagi RD (1979) Biotechnol Bioeng 21: 1387
32. Nelder JA, Mead R (1965) Computer J 7: 308
33. Shi Z, Shimizu K, Iijima S, Morisue T, Kobayashi T (1990) Biotechnol Bioeng 36: 520

Biological Nitrogen Removal from Wastewater

U. Wiesmann

Institut für Verfahrenstechnik, Technische Universität Berlin, Straße des 17. Juni 135, D-10623 Berlin, FRG

Advances in Biochemical Engineering
Biotechnology, Vol. 51
Managing Editor: A. Fiechter
© Springer-Verlag Berlin Heidelberg 1994

A review of the research on the kinetics of nitrification and denitrification is presented including an explanation of reaction engineering models. The results of laboratory scale experiments with high rate nitrification processes are discussed using kinetic results for oxygen limitation as well as for substrate limitation and inhibition. It can be demonstrated that reaction engineering models are helpful for a better understanding of the processes and for the design of reactors. Pilot scale investigations from the last 15 years show remarkable advances in the increase in nitrification efficiency and in the stabilization of the process. The time is ripe for nitrogen removal from industrial effluents in full scale processes!

List of Symbols

c	concentration	
c_B	bacteria concentration as odm (organic dry matter)	$\dfrac{M}{L^3}$
c'	concentration of dissolved oxygen	$\dfrac{M}{L^3}$
k_D	decay coefficient	$\dfrac{1}{T}$
k_e	endogenous respiration coefficient	$\dfrac{1}{T}$
K'	oxygen saturation coefficient	$\dfrac{M}{L^3}$
K_{SH}	saturation coefficient for the unionized substrate	$\dfrac{M}{L^3}$
K_{iH}	inhibition coefficient for the unionized substrate	$\dfrac{M}{L^3}$
K_S	saturation coefficient for the ionized substrate	$\dfrac{M}{L^3}$
K_i	inhibition coefficient for the ionized substrate	$\dfrac{M}{L^3}$
K_a, K_b	equilibrium constants for dissociation of the substrate	$\dfrac{M}{L^3}$
n_R	recycle ratio	–
n_E	thickening ratio	–
r	reaction rate	$\dfrac{M}{L^3T}$
r'	oxygen utilization rate	$\dfrac{M}{L^3T}$
r'^*	real maximal oxygen utilization rate	$\dfrac{M}{L^3T}$
r_{BW}	growth rate	$\dfrac{M}{L^3T}$

r_{Bd}	decay rate	$\dfrac{M}{L^3 T}$
$r_{O_2 S}$	oxygen utilization rate for substrate removal (e.g. ammonia)	$\dfrac{M}{L^3 T}$
$r_{O_2 e}$	oxygen utilization rate for endogenous respiration	$\dfrac{M}{L^3 T}$
S	Monod or Haldane term (Eq. 42)	–
T	temperature	°C
t	time	T
t_v	(hydraulic) mean residence time	T
t_{vB}	sludge age (mean residence time of bacteria)	T
μ	specific growth rate	$\dfrac{1}{T}$
μ_{max}	maximal specific growth rate	$\dfrac{1}{T}$
μ_{max}^*	real maximal specific growth rate	$\dfrac{1}{T}$

Indices

o	influent
a	effluent, reactor
M	mixing point
R	recycle flow
Ü	surplus sludge
k	critical
max	maximal

1 Introduction

Nitrogen is dissolved in many kinds of wastewaters as ammonia, nitrite and nitrate and as organic molecules such as amino acids. In addition it is contained in suspended organic particles. The mean content in sewage, for example, is about $40 \, \text{mg} \, l^{-1} \, NH_4 - N$ and about $20 \, \text{mg} \, l^{-1}$ org. N. During biological degradation the organic nitrogen is transformed to ammonia. In industrial effluents the ammonia concentration is often much higher (Table 1). Some industrial effluents are characterized by high nitrate concentrations (Table 2).

The following problems arise, if wastewater with ammonia and nitrate is discharged into rivers and lakes:

1. Ammonia is oxidized by bacteria to nitrite and nitrate leading to a decrease in the dissolved oxygen concentration and to a fish kill.

Table 1. Ammonia concentration and origin of some high-loaded wastewaters [17]

Author	Year	Ref.	Industry/Products	Conc. range N [mg l^{-1}]
Cousins WG, Mindler AB	1972	1	Cokery	3300–4100
Koziorowski B, Kucharski J	1972	2		800–1000
Adams CE, Eckenfelder WW	1977	3		450
Koziorowski B, Kucharski J	1972	2	Oil Refinery	23.8–752
Garrison WE et al.	1973	4		865
Pascik I	1982	5		450–630
Hutton WC, LaRocca SA	1975	6	Coal Gasification	< 1000
Koziorowski B, Kucharski J	1972	2		< 2500
Hutton WC, LaRocca SA	1975	6	Fertilizer	200–940
Arnold DW, Wolfram WE	1975	7		600
Hutton WC, LaRocca SA	1975	6	Synthetic Fibre	800
Meinck F et al.	1968	8	Slaughterhouse	145
ATV	1985	9	Livestock: Swine	2300
Braun R	1982	10	Livestock: Cattle	500–2300
Neumann H, Viehl K	1966	11	Rendering Plant	807
Zall RR	1972	12	Dairy	< 625
Basu AK	1975	13	Distillery	114–380
Patterson JW, Minear RA	1975	14	Explosives	< 1503
Adams CE, Eckenfelder WW	1977	3	Cellulose and Paper	264
US EPA,	1975	15	Glass	300–650
Brown GE	1975	16	Pharmaceuticals	475

Table 2. Nitrate concentration and origin of some extremely high-loaded wastewaters [22]

Author	Year	Ref.	Industry/Products	Concentration range N [mg l^{-1}]
Patterson JW	1958	18	Explosives	< 12 500
Francis CW Mankin JB	1977	19	Uranium processing	< 11 300
Patterson JW	1985	18	Fertilizer	< 6000
Jewell WJ Cummings RJ	1975	20	Electronic	< 2000
Bode H	1985	21	Pectin	< 2700

2. Ammonia and ammonium are in chemical equilibrium; with increasing temperature and pH more and more ammonia which is toxic to fish is produced.
3. Nitrate stimulates the growth of algae, contributing to the eutrophication of rivers and lakes.
4. High concentrations of nitrate and nitrite in drinking water cause methemoglobinemia in babies and favour the formation of cancerogenic nitrosamines.

For these reasons ammonia and nitrate have to be removed from discharges into rivers and lakes.

The minimum requirements for the discharge of municipal wastewater into inshore waters are laid down by the Frame Wastewater Regulation of the Federal Republic of Germany from 08.09.1989 [23]. The $NH_4 - N$-concentration of a two hour mixed sample must be lower than 10 mg l^{-1} for BOD_5-loads greater than 300 kg d^{-1}. That means, that all treatment plants with flow rates greater than 4000 m^3 d^{-1} have to be expanded by a nitrification step. The framework within which an expansion must be considered is given by the limit on the total inorganic nitrogen content (ammonia, nitrite and nitrate). In order to meet the limit of 18 mg N l^{-1}, a denitrification step must be added as well. The Frame Wastewater Regulation sets different limits for industrial effluents, ranging from 10 mg l^{-1} for the food industry to 100 mg l^{-1} $NH_4 - N$ for electroplating processes. The maximum discharge of nitrogen for the fertilizer industry is related to the mass of the product; for instance 4 kg per t $NH_4 - N$ is the limit for the production of a pure nitrogen fertilizer [24].

Therefore, it is evident that several billion DM must be invested in nitrogen removal both in municipal and industrial treatment plants.

2 Microbiology and Stoichiometry

2.1 Nitrification

Ammonia is oxidized to nitrite by chemolithoautotrophic bacteria. The most important strain is *Nitrosomonas europaea*, a short rod-like bacterium of

$1 \times 1.5 \, \mu m$ with a polar flagellum [25]. It follows from catabolism

$$NH_4^+ + \frac{3}{2}O_2 \rightarrow NO_2^- + H_2O + 2\,H^+ + 240 \div 350 \, kJ \qquad (1)$$

that a high amount of oxygen is needed and that the pH would decrease without pH control. From the stoichiometry for catabolism and anabolism [15]

$$55\,NH_4^+ + 76\,O_2 + 109\,HCO_3^- \rightarrow$$
$$C_5H_7NO_2 + 54\,NO_2^- + 57\,H_2O + 104\,H_2CO_3 \qquad (2)$$

the following real yield coefficients can be derived considering the consumption of bicarbonate by the produced protons and the composition of bacteria as $C_5H_7NO_2$[1]:

$$Y^o_{B/NH-N} = \frac{113}{14 \cdot 55} = 0.15 \text{ g odm per g } NH_4 - N\,^2$$

$$Y^o_{B/O_2} = \frac{113}{32 \cdot 76} = 0.047 \text{ g odm per g } O_2$$

$$Y^o_{O_2/NH_4-N} = \frac{76 \cdot 32}{55 \cdot 14} = 3.16 \text{ g } O_2 \text{ per g } NH_4 - N$$

$$Y_{HCO_3^-/NH_4-N} = \frac{109 \cdot 61}{55 \cdot 14} = 8.6 \text{ g } HCO_3^- \text{ per g } NH_4 - N.$$

The most important nitrite oxidizer is *Nitrobacter winogradskyi*, a short rod $(0.5 \times 1 \, \mu m)$ which can also appear as cocci [27], and which lives in symbiosis with Nitrosomonas.

The catabolism

$$NO_2^- + \frac{1}{2}O \rightarrow NO_3^- + 65 \div 90 \, kJ \qquad (3)$$

shows that less energy is available for growth in comparison to Nitrosomonas. Most authors use the following stoichiometry for catabolism and anabolism

$$400\,NO_2^- + NH_4^+ + 4H_2CO_3 + HCO_3^- + 195\,O_2 \rightarrow$$
$$C_5H_7NO_2 + 3\,H_2O + 400\,NO_3^- \qquad (4)$$

and consider that Nitrobacter needs $NH_4^+ - N$ for biosynthesis. Equation (4) gives the following real yield coefficients

$$Y^o_{B/NO_2-N} = \frac{113}{400 \cdot 14} = 0.02 \text{ g odm per g } NO_2 - N$$

---

[1] $M_{NH_4-N} = 14 \, g\,mol^{-1}$; $M_{O_2} = 32 \, g\,mol^{-1}$

$M_{C_5H_7NO_2} = 113 \, g\,mol^{-1}$; $M_{HCO_3^-} = 61 \, g\,mol^{-1}$

[2] odm = organic dry matter

$$Y_{B/O_2}^{\circ} = \frac{113}{195.32} = 0.018 \text{ g odm per g } O_2$$

$$Y_{O_2/NO_2-N}^{\circ} = \frac{195}{400} \frac{32}{14} = 1.11 \text{ g } O_2 \text{ per g } NO_2 - N.$$

Using the yield coefficients Y_{B/NH_4-N}° and Y_{B/NO_2-N}° oxidation of 1 kg $NH_4 - N$ would result in 150 g biomass of Nitrosomonas and 20 g biomass of Nitrobacter. Higher values for Y_{B/NO_2-N}° are frequently published in the literature resulting in an average value of

$$Y_{B/NO_2-N}^{\circ} = 0.042 \text{ g odm per g } NO_2 - N.$$

Most probably, the stoichiometry of Eq. (4) has to be corrected. The total real yield coefficient for the growth of nitrifiers is

$$Y_{B/NH_4-N}^{\circ} + Y_{B/NO_2-N}^{\circ} = 0.17 \text{ g odm per g } N$$

respectively

$$Y_{B/O_2}^{\circ} = 0.04 \text{ g odm per g } O_2$$

and lower than the real yield coefficient of aerobic heterotrophic bacteria:

$$Y_{B/O_2}^{\circ} = 0.72 \text{ g odm per g } O_2.$$

A typical characteristic of nitrification is a high oxygen consumption and a low biomass production!

2.2 Denitrification

Nitrate can be reduced in nature by the anabolism or catabolism of organisms. Only the catabolic use of nitrate is called denitrification. Denitrifiers can grow on CO_2 (autotrophs) and organics (heterotrophs). Denitrification by chemolithoautotrophic bacteria is used with success in the treatment of drinking water. Until now, only chemoorgano heterotrophic bacteria have been used in wastewater treatment.

A large number of species are able to use oxygen in an aerobic metabolism, and, in the absence of oxygen, to reduce nitrate in an anoxic metabolism (facultative aerobics). Therefore, the same biomass can be used in a combined aerobic/anoxic process for carbon and nitrate removal.

Nearly all denitrifiers are able to use NO_2^- instead of NO_3^- as an electron acceptor and a large number of different organics as electron donor or energy source. For example, for acetate as energy source the stoichiometry for catabolism is

$$5 CH_3COOH + 8 NO_3^- \rightarrow$$

$$4 N_2 + 10 CO_2 + 6 H_2O + 8 OH^- - 783 \text{ kJ mol}^{-1} \text{ acetate} \qquad (5)$$

or

$$3\,CH_3COOH + 8\,NO_2^- \rightarrow$$

$$4\,N_2 + 6\,CO_2 + 2\,H_2O + 8\,OH^-. \tag{6}$$

Because of the increase in pH, the inorganic carbon $(CO_2 + CO_3^{2-} + HCO_3^-)$ is mainly dissolved as CO_3^{2-}. Therefore, $CaCO_3$ is precipitated in the presence of Ca^{2+}. In most nitrogen removal processes denitrification is combined with nitrification. In the case of the treatment of industrial effluents with high ammonia concentration the amount of basic addition such as Na_2CO_3 or NaOH can be reduced for nitrification by the recycle of a $CaCO_3$ containing sludge from denitrification.

3 Kinetics

3.1 Apparent Yield Coefficients

The real yield coefficient for the growth of Nitrosomonas is

$$Y^o_{B/NH_4-N} = \frac{r_{BW}}{r_{NH_4-N}} \tag{7}$$

where r_{BW} = growth rate and r_{NH_4-N} = ammonia utilization rate.

If we consider the rate of bacteria decay by lysis and endogenous respiration r_{Bd}, we obtain the apparent yield coefficient

$$Y_{B/NH_4-N} = \frac{r_{BW} - r_{Bd}}{r_{NH_4-N}} \tag{8}$$

or

$$= Y^o_{B/NH_4-N}\left(1 - \frac{k_d}{\mu}\right) \tag{9}$$

with

$$r_{BW} = \mu\,c_B \tag{10}$$

and

$$r_{Bd} = k_d\,c_B \tag{11}$$

where μ = specific growth rate and k_d = decay coefficient.

For $\mu = \mu_{max} \gg k_d$ the real yield coefficient can be measured approximately. But in the case of substrate limitation μ can decrease considerably and a much lower apparent yield coefficient follows from Eq. (9). A further yield coefficient of similar importance is

$$Y^o_{O_2/NH_4-N} = \frac{r_{O_2S}}{r_{NH_4-N}} \qquad (12)$$

where r_{O_2S} = oxygen utilization rate for ammonia oxidation.

If we consider endogenous respiration r_{O_2e}, we get the apparent oxygen yield coefficient

$$Y_{O_2/NH_4-N} = \frac{r_{O_2S} + r_{O_2e}}{r_{NH_4-N}} \qquad (13)$$

and with

$$r_{O_2e} = \frac{k_e\, c_B}{Y_{B/O_2e}} \qquad (14)$$

and

$$r_{O_2S} = \frac{\mu\, c_B}{Y^o_{B/O_2}} \qquad (15)$$

$$Y_{O_2/NH_4-N} = Y^o_{B/NH_4-N} + \frac{k_e}{\mu}\,\frac{1}{Y_{B/O_2e}}. \qquad (16)$$

Y_{B/O_2e} is the yield for biomass loss during endogenous respiration. For low specific growth rate resulting from substrate limitation or inhibition, the apparent yield coefficient can be considerably higher than the real coefficient.

Equations (9) and (16) are valid for Nitrobacter and for all other aerobic or anoxic bacteria.

3.2 Nitrification

Ammonia in its unionized form is the real electron donor (substrate) for Nitrosomonas, because less energy is required for its transport into the cell in comparison to the transport of an ionized molecule like NH_4^-.

Substrate inhibition has often been described in the literature of the last decade. The simplest description of substrate inhibition is given by Haldane-kinetics [28]:

$$\mu = \mu_{max} \frac{c(NH_3 - N)}{K_{SH} + c(NH_3 - N) + \dfrac{c(NH_3 - N)^2}{K_{iH}}}\,\frac{c'}{K' + c'}. \qquad (17)$$

The second factor in Eq. (17) describes oxygen-limitation. For $c' \gg K'$ and $K_{iH} \gg c(NH_3 - N)^2$ Eq. (17) becomes

$$\mu = \mu_{max} \frac{c(NH_3 - N)}{K_{SH} + c(NH_3N)}. \qquad (18)$$

Ammonia inhibition can be always neglected for $pH < 9$, $c(N) < 65 \, mg \, l^{-1}$ ($NH_4 - N + NH_3 + N$) and therefore in all sewage treatment plants!

From

$$NH_4^+ \rightarrow NH_3 + H^+ \tag{19}$$

we get

$$c(NH_3 - N) = \frac{c(NH_4 - N) \times 10^{pH}}{K_a} \tag{20}$$

where

$$K_a = \exp\left(\frac{6344}{273 + T}\right) \tag{21}$$

as the equilibrium constant.

Equations (20) and (18) give

$$\mu = \mu_{max} \frac{c(NH_4 - N)}{K_S + c(NH_4 - N)} \tag{22}$$

and

$$K_S = K_{SH} K_a 10^{-pH}. \tag{23}$$

Equation (22) is used in all kinetic studies with low ammonia concentrations and in models for nitrification in sewage treatment plants. In contrast, for the treatment of many industrial effluents

$$\mu = \mu_{max} \frac{c(NH_3 - N)}{K_{SH} + c(NH_3 - N) + \dfrac{c(NH_3 - N)^2}{K_{iH}}} \tag{24}$$

must be used, or, in the case of simultaneous oxygen limitation Eq. (17).

There is a similar influence of the unionized electron donor HNO_2 in the growth rate of Nitrobacter:

$$\mu = \mu_{max} \frac{c(HNO_2 - N)}{K_{SH} + c(HNO_2 - N) + \dfrac{c(HNO_2 - N)^2}{K_{iH}}} \frac{c'}{K' + c'} \tag{25}$$

where

$$c(HNO_2 - N) = \frac{c(NO_2^- - N)}{K_b \times 10^{pH}} \tag{26}$$

and

$$K_b = \exp\left(\frac{-2300}{273 + T}\right) \tag{27}$$

the equilibrium constant for

$$NO_2^- + H^+ \rightarrow HNO_2. \tag{28}$$

For the case of ammonia oxidation in sewage treatment plants, substrate inhibition can be neglected, resulting in

$$\mu = \mu_{max}\frac{c(NO_2 - N)}{K_S + c(NO_2 - N)} \qquad (29)$$

for oxygen concentrations $c' \gg K'$. In most of the models of nitrification in sewage treatment plants ammonia oxidation is assumed to be the slowest step. Therefore the influence of nitrite oxidation rate on nitrate formation is often neglected. We will later demonstrate that this is not allowable for higher loaded industrial effluents!

3.3 Denitrification

The specific growth rate is influenced by both the concentration of the organic substrate and the concentration of the electron acceptor NO_3^- or NO_2^-.

In most papers a double Monod-Kinetic is used, which gives, in the case of NO_3^- as electron acceptor,

$$\mu = \mu_{max}\frac{c_S}{K_S + c_S}\frac{c(NO_3 - N)}{K_{NO_3} + c(NO_3 - N)} \qquad (30)$$

and for NO_2^-

$$\mu = \mu_{max}\frac{c_S}{K_S + c_S}\frac{c(NO_2 - N)}{K_{NO_2} + c(NO_2 - N)}. \qquad (31)$$

There are differences in the maximal growth rates μ_{max} and in the saturation coefficients K_S and K_{NO_2} because of the lower reduction equivalent of NO_2^- but normally it is the same bacteria using either NO_3^- or NO_2^-! From Eq. (30) the substrate utilization and the denitrification rates can be calculated by

$$r_S = \frac{\mu}{Y_{B/S}^o}c_B \qquad (32)$$

and

$$r_{NO_3-N} = \frac{\mu}{Y_{B/NO_3-N}^o}c_B = \frac{1}{2}r_{N_2}. \qquad (33)$$

Corresponding equations follow from Eq. (31).

3.4 Examples for the Measurement of Kinetic Coefficients

3.4.1 Inhibition Coefficients K_{iH} for Nitrobacter and Nitrosomonas

The direct test of Eq. (17) and the estimation of K_{iH} is very difficult because of the required measurement of

$$r_{BW} = \mu c_{NS} \qquad (32)$$

and c_{NS} the biomass concentration of Nitrosomonas. It is advantageous to measure the oxygen consumption rate using

$$r' = \frac{r_{BW}}{Y^0_{NS/O_2}} \tag{33}$$

instead of Eq. (32).

From Eq. (17), it follows that for relatively high oxygen concentration ($c' \gg K'$), neglecting endogenous respiration

$$r' = r_{O_2/NS} = \frac{\mu_{max}}{Y^0_{NS/O_2}} \frac{c(NH_3 - N)}{K_{SH} + c(NH_3 - N) + \dfrac{c(NH_3 - N)^2}{K_{iH}}} c_{NS}. \tag{34}$$

For Haldane-kinetics the maximal growth rate μ_{max} cannot be realized. Before substrate limitation is finished, if $c(NH_3 - N)$ is increased, substrate inhibition will start. The real maximal growth rate

$$\mu^*_{max} = \frac{\mu_{max}}{1 + 2\sqrt{\dfrac{K_{SH}}{K_{iH}}}} \tag{35}$$

follows from Eq. (17) by the determination of the maximum. Only for $K_{iH} \gg K_{SH}$ can the theoretical maximal growth rate be reached approximately. The corresponding real maximal oxygen consumption rate is

$$r'^* = \frac{\mu^*_{max}}{Y^0_{NS/O_2}} c_{NS}. \tag{36}$$

If we divide Eq. (34) by (36) considering (35) we obtain

$$\frac{r'}{r'^*} = \frac{c(NH_3 - N)}{K_{SH} + c(NH_3 - N) + \dfrac{c(NH_3 - N)^2}{K_{iH}}} \left(1 + 2\sqrt{\frac{K_{SH}}{K_{iH}}}\right). \tag{37}$$

A corresponding relation follows for nitrite oxidation:

$$\frac{r'}{r'^*} = \frac{c(HNO_2 - N)}{K_{SH} + c(HNO_2 - N) + \dfrac{c(HNO_2 - N)^2}{K_{iH}}} \left(1 + 2\sqrt{\frac{K_{SH}}{K_{iH}}}\right). \tag{38}$$

In order to validate the theory described above, measurements of the kinetic coefficients were carried out in a batch reactor. The mixed bacteria culture containing Nitrosomonas and Nitrobacter was removed from continuous lab. scale nitrification processes with $(NH_4)_2SO_4$ as the only substrate. In the investigation of nitrite oxidation kinetics, $NaNO_2$ and biomass were added to a batch reactor containing a solution of nutrients in tap water. In spite of the presence of Nitrosomonas oxygen is only removed by Nitrobacter, if endogenous respiration is neglected.

For nearly constant concentrations of $c(HNO_2)$ and biomass, r' follows from the oxygen balance

$$\frac{dc'}{dt} = \frac{\Delta c'}{\Delta t} = r' \qquad (39)$$

if $c' = f(t)$ is a straight line. The results plotted in Fig. 1 as r'/r'^* versus $\ln c(HNO_2 - N)$ for a large number of different nitrite concentrations and pH-values show an increase in the oxygen consumption rate at low concentrations and a decrease in the region of higher concentrations. The coefficients K_{SH} and K_{iH} can be estimated by linearization of Eq. (38), each at low concentrations by neglecting limitation and at high concentrations neglecting inhibition [17].

The solid line in Fig. 1 was calculated with Eq. (38) using the estimated coefficients

$$K_{SH} = 3.9 \times 10^{-5} \text{ mg l}^{-1} \text{ HNO}_2 - N$$

$$K_{iH} = 0.26 \text{ mg l}^{-1} \text{ HNO}_2 - N.$$

It shows good agreement with the experimental results. The nitrifiers came from three continuous lab. reactors, one fluidized bed and two stirred tank reactors.

Nearly the same procedure was repeated for measuring the ammonia oxidation $(NH_4)_2SO_4$ instead of $NaNO_2$. But, now oxygen is consumed by both Nitrosomonas and Nitrobacter, because it was not possible to find a substance, which would inhibit only Nitrobacter. Therefore, all data have to be corrected by substracting the known amount of oxygen consumed by nitrite oxidation.

The results for a large number of different concentrations of ammonium ions and different pH-values are plotted in Fig. 2a as r'/r'^* versus $c(NH_4 - N)$. The considerable scatter of the data can be remarkably reduced by plotting the

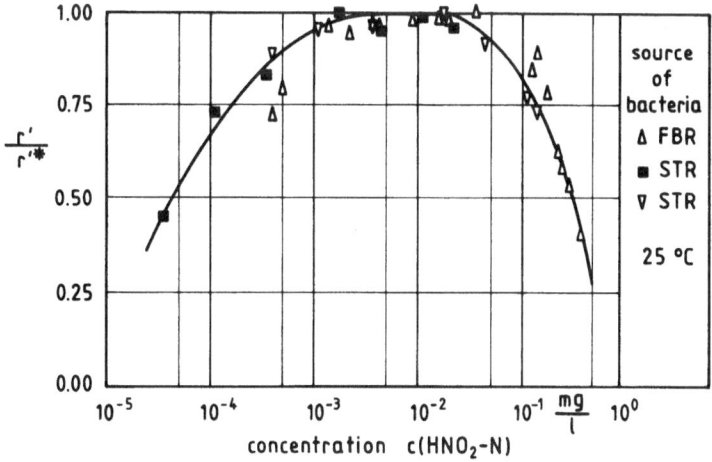

Fig. 1. Substrate limitation and inhibition by $HNO_2 - N$. Solid line: Eq. (38) with $K_{SH} = 3.9 \times 10^{-5}$, $K_{iH} = 0.26 \text{ mg l}^{-1} HNO_2 - N$

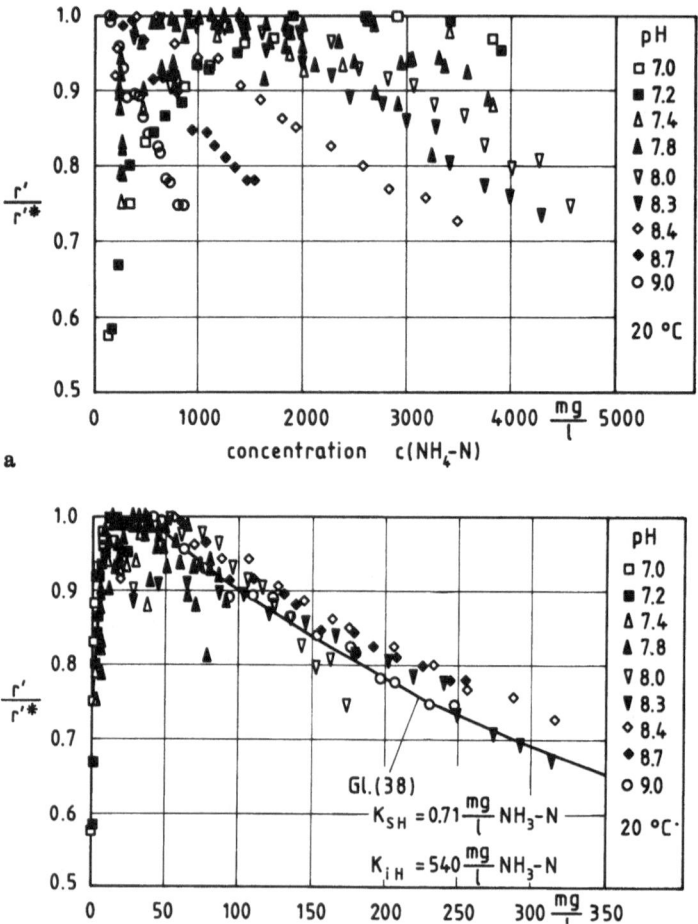

Fig. 2. a) Substrate limitation and inhibition of Nitrosomonas by ammonium ions. **b)** Substrate limitation and inhibition of Nitrosomonas by ammonia

results against $c(NH_3 - N)$ (Fig. 2b) [17], confirming the properties of ammonia as an inhibitor substrate. The results can be fitted very well by Eq. (37) and the coefficients

$$K_{SH} = 0.71 \text{ mg l}^{-1} \text{ NH}_3 - N$$

and

$$K_{iH} = 540 \text{ mg l}^{-1} \text{ NH}_3 - N$$

are not influenced by pH.

As follows from Eq. (35) 93% of the theoretical maximal growth rate can be reached! If $NH_4 - N$ or $NO_2 - N$ are to be used as substrate concentrations, the pH dependent K_s and K_i values can be calculated by Eq. (22).

3.4.2 Oxygen Saturation Coefficient K' for Nitrobacter

Corresponding to Eq. (34) we can write for the oxygen consumption rate of Nitrobacter

$$r' = r_{O_2/Nb}$$

$$= \frac{\mu_{max}}{Y^o_{Nb/O_2}} \frac{c(HNO_2 - N)}{K_{SH} + c(HNO_2 + N) + \dfrac{c(HNO_2 - N)^2}{K_{iH}}} \left[\frac{c'}{K' + c'}\right] c_{Nb} \quad (40)$$

if oxygen limitation is considered.

For the investigation of oxygen limitation HNO_2-concentration should be nearly be constant giving

$$r' = r_{O_2/Nb} = \frac{\mu_{max}}{Y^o_{Nb/O_2}} \frac{S}{} \frac{c'}{K' + c'} c_{Nb} \quad (41)$$

with

$$S = \frac{c(HNO_2 - N)}{K_{SH} + c(HNO_2 - N) + \dfrac{c(HNO_2 + N)^2}{K_{iH}}} = const. \quad (42)$$

The oxygen balance for a batch reactor with decreasing oxygen concentration is

$$\frac{dc'}{dt} = r'_{max} S \frac{c'}{K' + c'} \quad (43)$$

with

$$r'_{max} = \frac{\mu_{max} c_{Nb}}{Y^o_{Nb/O_2}}. \quad (44)$$

After integration and linearization we obtain

$$\frac{c'_0 - c'}{\ln \dfrac{c'_0}{c'}} = r'_{max} S \frac{t}{\ln \dfrac{c'_0}{c'}} - K'. \quad (45)$$

Figure 3a represents the decrease of oxygen concentration with time for $c(NO_2 - N) = 50$ mg l^{-1}, pH = 7.8 and T = 25 °C. The concentration of Nitrobacter is unknown and not necessary for the test of Eq. (43) and the determination of K'.

In Fig. 3b the same data is plotted corresponding to Eq. (45) showing a straight line with an intersection with the ordinate of

$$K' = 1.09 \text{ mg } l^{-1} O_2.$$

Thus, the usefulness of the Monod-kinetic formulation for the description of oxygen limitation could be confirmed [17].

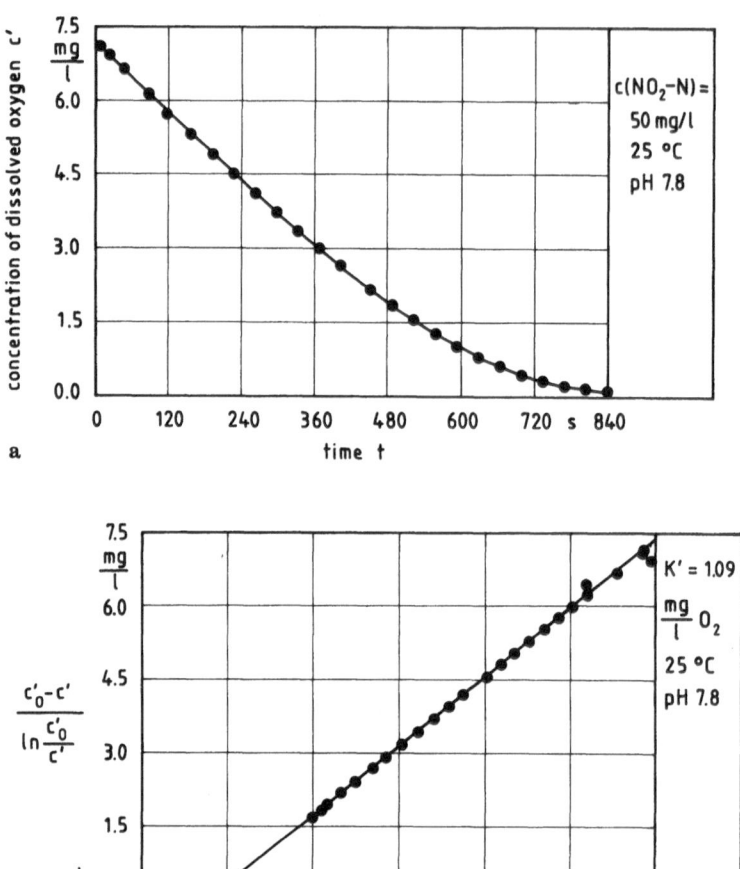

Fig. 3. a) Oxygen respiration by nitritoxidation. **b)** Linearization of data corresponding to Eq. (45)

3.4.3 Substrate Saturation Coefficient for Denitrification of Nitrate

Running similar experiments, the substrate limitation of denitrification was studied. The biomass was removed from a lab scale continuous stirred tank reactor with sludge recycle after sedimentation [22]. For a high nitrate concentration no nitrate inhibition would occur during batch experiment, which can be described by

$$\frac{dc_s}{dt} = -\frac{\mu_{max}}{Y^o_{B/NO_3}} \frac{c_s}{K_s + c_s} c_B \tag{46}$$

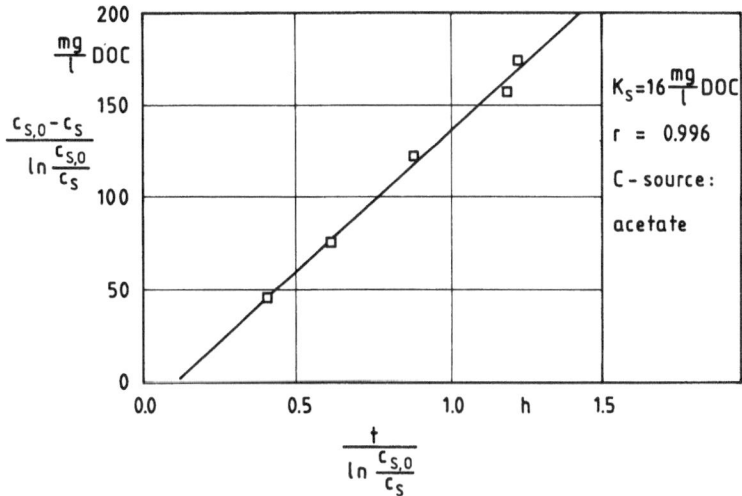

Fig. 4. Acetate consumption by denitrification with nitrate, linearization of data corresponding to the integral of Eq. (47)

$$\frac{dc_s}{dt} = - r_{smax} \frac{c_s}{K_S + c_s}.$$ (47)

After integration and linearization the function can be written in the form of Eq. (45).

The experiment was carried out with acetate as the energy and carbon source. After sampling, its concentration was measured by gas chromatography. Figure 4 shows a straight line with

$$K_s = 16 \text{ mg } l^{-1} \text{ DOC}$$

confirming the usefulness of Monod-kinetics [22].

3.5 Kinetic coefficients

Table 3 presents kinetic coefficients for nitrification, denitrification and aerobic degradation of organic compounds. The coefficients are the averaged values of widely scattered data collected from literature [33, 34] or recent measurements [17, 22, 35]. They can be used in reaction engineering models if only approximate results are needed. It may be necessary to make own kinetic measurements especially for the degradation of mixed organic substrate if more exact results are desirable.

With the exception of real yield coefficients, all other coefficients are dependent on temperature. In some cases this dependence has been studied experimentally (Table 4).

Table 3. Kinetic coefficients for $T = 20\,°C$: 1) data for pH 8 – typical data from literature and own measurements, odm = organic dry matter, $25\ g\ COD = 1\ g\ DOC$

NH$_4$-oxidation	NO$_2$-oxidation	Aerobic degradation of organics	NO$_3$-reduction	NO$_2$-reduction
$\mu_{max} = 0.77\ d^{-1}$	$\mu_{max} = 1.08\ d^{-1}$	$\mu_{max} = 7.2\ d^{-1}$	$\mu_{max} = 2.6\ d^{-1}$	$\mu_{max} = 1.5\ d^{-1}$
$Y^\circ_{B/NH_4-N} = 0.147\,\dfrac{g\ odm}{g\ NH_4-N}$	$Y^\circ_{B/NO_2-N} = 0.042\,\dfrac{g\ odm}{g\ NO_2-N}$	$Y^\circ_{B/S} = 0.43\,\dfrac{g\ odm}{g\ COD}$	$Y^\circ_{B/S} = 0.4\,\dfrac{g\ odm}{g\ COD}$	$Y^\circ_{B/S} = 0.4\,\dfrac{g\ odm}{g\ COD}$
$k_d = 0.048\ d^{-1}$	$k_d = 0.048\ d^{-1}$	$k_d' = 0.24\ d^{-1}$	$k_d = 0.1\ d^{-1}$	$k_d = 0.1\ d^{-1}$
$k_e = 0.005\ d^{-1}$	$k_e = 0.005\ d^{-1}$	$k_e = 0.19\ d^{-1}$		
$Y^\circ_{O_2/NH_4-N} = 3.16\,\dfrac{g\ O_2}{g\ NH_4-N}$	$Y^\circ_{O_2/NO_2-N} = 1.1\,\dfrac{g\ O_2}{g\ NO_2-N}$	$Y^\circ_{O_2/S} = 0.6\,\dfrac{g\ O_2}{g\ COD}$	$Y^\circ_{NO_3/S} = 0.33\,\dfrac{g\ NO_3-N}{g\ COD}$	$Y^\circ_{NO_2/S} = 0.53\,\dfrac{g\ NO_2-N}{g\ COD}$
$K_S^1 = 0.7\ mg\,l^{-1}\ NH_4-N$	$K_S^1 = 1.3\ mg\,l^{-1}\ NH_4-N$	$K_S = 100\ mg\,l^{-1}\ COD$	$K_S = 25\ mg\,l^{-1}\ COD$	
$K_I^1 = 13\,500\ mg\,l^{-1}\ NH_4-N$	$K_I^1 = 10\,400\ mg\,l^{-1}\ NH_4-N$			
$K_{SH} = 2.8\,10^{-2}\ mg\,l^{-1}\ NH_3-N$	$K_{SH} = 3.2\,10^{-5}\ mg\,l^{-1}\ HNO_2-N$	$Y^\circ_{B/O_2} = 0.72\,\dfrac{g\ odm}{g\ O_2}$	$Y^\circ_{B/NO_3} = 1.2\,\dfrac{g\ odm}{g\ NO_3-N}$	$Y^\circ_{B/NO_2} = 0.8\,\dfrac{g\ odm}{g\ NO_2-N}$
$K_{IH} = 540\ mg\,l^{-1}\ NH_3-N$	$K_{IH} = 0.26\ mg\,l^{-1}\ HNO_2-N$			
$K' = 0.3\ mg\,l^{-1}\ O_2$	$K' = 1.1\ mg\,l^{-1}\ O_2$	$K' = 0.08\ mg\,l^{-1}\ O_2$	$K_{NO_3} \le 0.14\ mg\,l^{-1}\ NO_3-N$	$K_{NO_2} \le 0.12\ mg\,l^{-1}\ NO_2-N$

Table 4. Temperature dependence of some measured kinetic coefficients; T in °C, pH 8

Process	Coefficient	Dependence on temperature	Data range mean value for T = 20°C		Author	Ref.
NH$_4$-oxidation by	μ_{max}	$\exp(0.0951\,T - 2.174)$	$8 < T < 30°C$	0.76	Knowles et al. 1965	29
Nitrosomonas	d^{-1}	$0.29\,\exp(0.11\,(T - 10))$	$6 < T < 14°C$		Gujer 1977	30
		$\mu_{15°c}\,\exp(0.95(T - 15))$			Gray 1989	31
	K_S	$\exp(0.1174\,T - 2.666)$	$8 < T < 30°C$	0.72	Knowles et al. 1965	29
	$\dfrac{mg}{l}\,NH_4 - N$	$0.405\,\exp(0.118\,(T - 15))$	$5 < T < 30°C$	0.42	Stankevich 1972	32
	$\dfrac{1}{K'}\,\dfrac{mg}{l}\,O_2$	$0.21\,\exp(0.069\,(T - 15))$	$5 < T < 30°C$	0.43		
NO$_2$-oxidation	μ_{max}	$\exp(0.0587T - 1.13)$	$8 < T < 30°C$	1.04	Knowles et al. 1965	29
by Nitrobacter	d^{-1}	$0.79\,\exp(0.069(T - 15))$			Jenkins 1969	30
	K_S	$\exp(0.145T - 2.646)$	$8 < T < 30°C$	1.29	Knowles et al. 1965	29
	$mgNO_2 - N$	$0.625\,\exp(0.146(T - 15))$	$5 < T < 30°C$	1.30	Stankevich 1972	32
aerobic degra- dation of organic matter	μ_{max} d^{-1}	$1.37 \times 10^3\,\exp\left(-\dfrac{4222}{T + 273}\right)$	$10 < T < 70°C$	7.2	Wiesmann 1986	33

Some typical qualities of microbial systems can be discussed by using the given kinetic coefficients:

a) Composition of Autotrophic Nitrifying Biomass

Assuming an unlimited growth of nitrifiers and no accumulation of nitrite, the composition of the biomass, i.e. the ratio of Nitrosomonas to Nitrobacter, can be calculated by

$$r_{NH_4-N} = \frac{\mu_{max,NS}\, c_{NS}}{Y^o_{B/NH_4-N}} = r_{NO_2-N} = \frac{\mu_{max,Nb}\, c_{NS}}{Y^o_{B/NO_2-N}}$$

$$\frac{c_{Nb}}{c_{NS}} = \frac{0.77}{1.08}\frac{0.042}{0.147} = 0.20$$

$$\frac{c_{Nb}}{c_B} \times 100 = 17\% \quad \text{for} \quad c_B = c_{NS} + c_{Nb}. \tag{48}$$

This result is often used in evaluating the kinetics of nitrification, because differentiation of the two species using microbiological methods is difficult and time-consuming.

b) Oxygen Limitation of Growing Heterotrophic and Autotrophic Bacteria

If substrate limitation is avoided in an aerobic system, oxygen limitation can be described using Monod-kinetics

$$\mu = \mu_{max}\frac{c'}{K' + c'}. \tag{49}$$

For an oxygen concentration of $c' = 1\ mg\,l^{-1}\ O_2$ the relative specific growth rate is given by

$$\frac{\mu}{\mu_{max}} = \frac{1}{K' + 1} \tag{50}$$

and follows from the different K'-values of Table 3:

aerobic degradation of organics: $\dfrac{\mu}{\mu_{max}} = 0.93$

NH$_4$-oxidation $\dfrac{\mu}{\mu_{max}} = 0.77$

NO$_2$-oxidation $\dfrac{\mu}{\mu_{max}} = 0.48.$

If we want to ensure a relative specific growth rate of Nitrobacter of $\mu/\mu_{max} = 0.93$, an oxygen concentration of

$$c' = \frac{0.93}{0.07} K' = 13.3 \text{ mg } l^{-1} O_2$$

is needed, which can be obtained only by using pure oxygen instead of air!

c) Elimination of Organic Compounds in Sewage by Total Denitrification

If sewage with 300 mg l^{-1} COD and 40 mg l^{-1} $NH_4 - N$ is to be treated by carbon removal, nitrification and denitrification, (neglecting endogenous respiration and bacteria lysis and assuming complete nitrification and denitrification) the removed concentration of organics follows (Table 3) from

$$Y^0_{NO_3/S} = \frac{c(NO_3 - N)}{c_s} = 0.33 \text{ g } NO_3 - N \text{ per g COD}$$

to

$$c_s = \frac{c(NO_3 - N)}{Y^0_{NO_3/S}} = 121 \text{ mg } l^{-1} \text{ COD}.$$

Therefore, $121/300 \times 100 = 40\%$ of the costs for aeration, needed for the aerobic degradation of organics can be saved if nitrification is coupled with denitrification in a sewage treatment plant.

3.6 Inhibition of Nitrosomonas and Nitrobacter

Nitrosomonas is inhibited by the product NO_2^- or HNO_2 and some authors have tried to describe it as a competitive or non-competitive inhibition and to determine inhibition coefficients [36–38]. Because of the high scattering of the data [17] the coefficients are not published here.

Thiourea or allylthiourea is used as an inhibitor of nitrification in BOD_5-measurements. In mixed cultures (i.e. in biological treatment units) the concentration of a toxic compound causing inhibition is much higher than that in pure culture studies. In the pure culture, the organism is completely inhibited at 0.5 mg l^{-1} thiourea. However, in activated sludge up to 92 mg l^{-1} thiourea can be tolerated by Nitrosomonas. Obviously, nitrifying bacteria are able to acclimatize to relatively high inhibitor concentrations. Therefore, they are rarely inhibited by the range of ions found under normal wastewater conditions [31]. A large number of inhibitors were compiled by Richardson [39] and Bédard and Knowles [40].

3.7 Nitrification in Biofilms

Nitrifiers grow on the surfaces of solid particles, such as sand, forming biofilms. Particles with diameters > 2 mm are used in fixed bed reactors with direct aeration at the bottom of the reactor or in a fluid recycle stream. Smaller particles with diameters < 0.5 mm are used in fluidized bed reactors resulting in a higher biomass for the same biofilm thickness.

Because of the relative high K'-values of Nitrobacter and Nitrosomonas (see Sect. 3.5), oxygen limitation cannot be avoided in biofilms. Therefore, all reaction rates are influenced by diffusion and the biofilm thickness. Nitrification in biofilms is described by the effectiveness factor, which can be calculated by considering diffusion terms in the balances of oxygen, ammonia and nitrite [41–43]. Substrate inhibition following Eqs. (17) and (25) as well as mass transfer resistence at solid and bubble surfaces was considered for the first time by Larsen-Vefring [44]. The models must be improved by considering H^+-balances, in order to determine the local pH value and the real concentration of ammonia and nitrous acid.

4 Bioreactor Systems for Nitrification and Denitrification

4.1 Municipal Treatment Plants

Typical of municipal wastewater is a COD of 300 mg l^{-1}, a $NH_4 - N$ concentration of 40 mg l^{-1} and a $NO_3 - N$ concentration of 0 mg l^{-1}. Nevertheless, three problems have to be solved: carbon removal, nitrification and denitrification for relatively high flow rates up to 300 000 $m^3 d^{-1}$ in an economic way. A one step process can be realized in an oval oxidation ditch with extended aeration using brush aerators which give a horizontal flow for the needed recirculation (Fig. 5a). Due to the low sludge load of 0.03–0.15 kg BOD_5 kg $dm^{-1} d^{-1}$, or a high sludge age of > 10 d the nitrifiers are not washed out, so that nitrification occurs simultaneously with aerobic carbon removal. For denitrification an oxygen deficiency is needed, which can be achieved by a limited aeration and a sufficient distance between two rotors. So, oxic and anoxic zones alternate resulting in denitrification of the formed nitrates. The process must be controlled by changing the speed or the depth of the submerged part of the aerator. In large treatment plants the anoxic zone is often separated from the aerobic zone. Two different processes are possible, an anoxic stage followed by an aerobic one (Fig. 5b) or an aerobic stage followed by an aerobic one (Fig. 5c). In the first case, part of the nitrified wastewater must be recycled to the anoxic stage in order to supply the facultative anaerobic denitrifying bacteria with the electron acceptor nitrate. This type of process is used frequently. In the second case, part of the raw water must be fed to the denitrification stage in

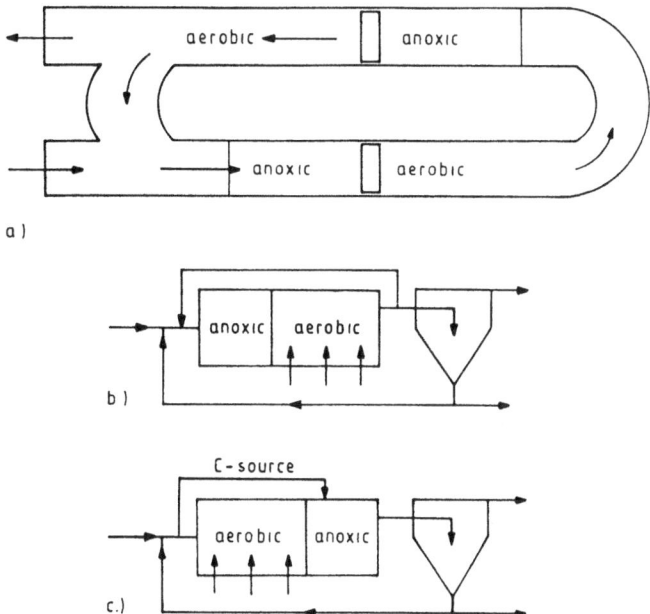

Fig. 5a–c. Municipal wastewater treatment with nitrification and denitrification. **a)** Simultaneous nitrification and denitrification. **b)** Denitrification followed by nitrification. **c)** Nitrification followed by denitrification

order to supply the denitrifying bacteria with organics as electron donors. Because of the low BOD_5-concentration required in the effluent, this raw water flow rate must be controlled using on-line measurements of the nitrate concentration in the influent of the anoxic stage. Instead of activated sludge processes, trickling filters and other attached growth processes can be used for aerobic carbon removal and nitrification. For denitrification the contact reactor must be submerged.

4.2 Treatment of Industrial Effluents

Three different types of wastewater must be distinguished for the planning of the process:
a) Wastewater with nitrate or nitrate and organics
b) Wastewater with ammonia or ammonia and nitrate
c) Wastewater with high concentrations of organics and ammonia.

a) Wastewater with Nitrate or Nitrate and Organics

If nitrate is the only component, that has to be removed, a carbon source must be added. Normally a wastewater with a low concentration of ammonia and organic nitrogen is preferred in order to avoid toxicity problems with ammonia ions. The treatment is carried out in a one step process. Organics (for example methanol, acetic acid or a wastewater) is added proportional to the nitrate concentration. Different kinds of reactor types can be used e.g. activated sludge systems, submersed disc contactors, fixed bed, and fluidized bed reactors. A two stage activated sludge process with an anoxic first stage and an aerobic second stage with sludge recycle is a conventional process for the treatment of wastewater with nitrate and organics.

b) *Wastewater with Ammonia or Ammonia and Nitrate*

Nitrification occurs in the aerobic first stage and denitrification in the anoxic second stage, after organics are added. Besides the activated sludge system, all kinds of attached growth systems such as rotating disc contactors (for denitrification submerged rotating disc contactors), fixed bed reactors, and fluidized bed reactors can be used. If flocs with a relatively high settling rate are produced, fluidized bed reactors without solid support particles have proved successful.

c) *Wastewater with High Concentrations of Organics and Ammonia*

Problems with oxygen limitation of Nitrobacter and Nitrosomonas can be better avoided in a process with separate biomass recycle systems (Fig. 6a). In

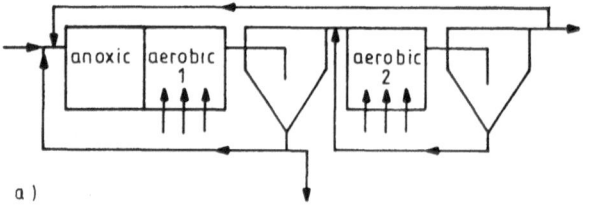

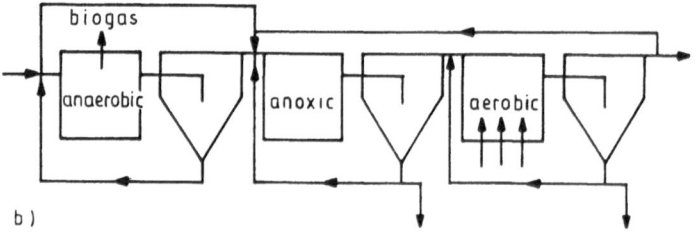

Fig. 6a, b. Treatment of industrial effluents by carbon removal, nitrification and denitrification. **a)** Two-sludge system with aerobic carbon removal. **b)** Three-sludge system with anaerobic carbon removal

the first activated sludge system consisting of an anoxic and an aerobic stage with a sedimentation tank and sludge recycle, only heterotrophic bacteria can grow if the sludge age is short enough, causing a washout of the nitrifiers. In the second activated sludge system nitrifiers predominate. A high concentration of nitrifiers can be established, because no surplus sludge must be removed due to the low yield coefficients $Y^0_{B/N}$ (see Table 3). Normally the bacteria concentration in the overflow of the second sedimentation tank can be tolerated. In order to keep the heterotrophic and autotrophic biomass separate, sedimentation must precede the recycle of the nitrate-containing water to the first stage. If 80% of the treated water is recycled to the anoxic stage, the hydraulic load of the settling tanks is increased by a factor of 5, resulting in a corresponding increase in settling area. Such large settling tanks can be avoided if attached growth systems are used instead of activated sludge systems. Presumably, the development will go in this direction!

For wastewaters with a high content of organics, anaerobic treatment should be preferred because of the lower amount of surplus sludge produced and the savings in energy costs for aeration. In addition, the energy of the biogas with a content of 60 to 80% CH_4 can be used partly for other purposes than the heating of the anaerobic stage.

Figure 6b shows a three-sludge system with an anaerobic first stage, an anoxic second stage and an aerobic third stage, all with settling tanks and separate sludge recycling systems. The anaerobic process must be controlled to produce a concentration of easily assimilable organics, mainly lower fatty acids produced by acidification, high enough for denitrification.

As discussed previously, the process can be improved by using attached growth reactors. For pH control of nitrification, basic additives such as Na_2CO_3, $CaCO_3$ or NaOH must be used. The cations Na^+ and Ca^{2+} partly recycled into the anoxic stage, are precipitated and removed with the surplus sludge. Here, an advantage of the one-sludge system is obvious (Fig. 5): the precipitated salts are transferred to the nitrification stage resulting in a saving of chemicals for pH control!

5 Reaction Engineering Modelling of the Nitrification Process in Municipal Wastewater Treatment Plants

The design of municipal wastewater treatment plants is carried out in the Federal Republic of Germany using the guidelines A 131 from the ATV (Abwassertechnische Vereinigung, 1991) [45]. A better understanding of the process with regard to biokinetics follows reaction engineering models. The method will be demonstrated for nitrification of sewage by an activated sludge process in a completely stirred tank reactor (CSTR) for steady state conditions. An outline of the process with concentration profiles is given in Fig. 7

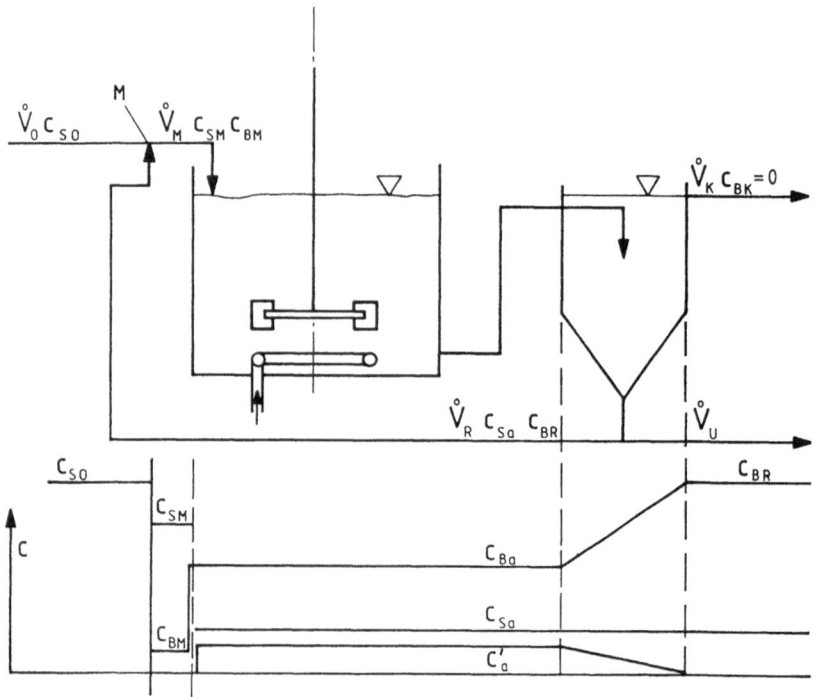

Fig. 7. Nitrification in a CSTR with sedimentation and sludge return

It is assumed that no oxygen limitation occurs and no nitrite accumulates. Under sewage treatment conditions (pH < 9, $c(N) < 65 \text{ mg l}^{-1}$ ($NH_4 - N$ + $NH_3 - N$), see Sect. 3.2) ammonia inhibition can be neglected. Therefore Eq. (22) can be used for the specific growth rate of Nitrosomonas.

The model consists of the balances for $NH_4 - N$

$$0 = \dot{V}_M(c_{SM} - c_{Sa}) - \frac{\mu_{max} c_{Sa} c_{Ba}}{Y^o_{B/NH_4-N}(K_S + c_{Sa})} V_R \tag{51}$$

and Nitrosomonas

$$0 = \dot{V}_M(c_{BM} - c_{Ba}) + \left(\frac{\mu_{max} c_{Sa} c_{Ba}}{K_S + c_{Sa}} - k_d c_{Ba} \right) V_R \tag{52}$$

where c_s = concentration of $NH_4 - N$ and c_B = concentration of Nitrosomonas.

From balances on the mixing point M we obtain

$$\dot{V}_M = \dot{V}_o + \dot{V}_R \tag{53}$$

$$c_{SM} = \frac{c_{So} + n_R c_{Sa}}{1 + n_R} \tag{54}$$

$$c_{BM} = \frac{n_R n_E c_{Ba}}{1 + n_R} \tag{55}$$

where

$$n_R = \dot{V}_R / \dot{V}_o = \text{recycle ratio} \tag{56}$$

$$n_E = c_{Br} / c_{Ba} = \text{thickening ratio}. \tag{57}$$

From Eqs. (51)–(57) the $NH_4 - N$ concentration of the effluent follows

$$c_{Sa} = \frac{K_S k_d t_v + (1 + n_R - n_E n_R) K_S}{t_v(\mu_{max} - k_d) - (1 + n_R - n_E n_R)} \tag{58}$$

where

$$t_v = \frac{V_R}{\dot{V}_o} = \text{mean residence time of the water}. \tag{59}$$

A more useable parameter of the activated sludge process is

$$t_{vB} = \frac{V_R c_{Ba}}{\dot{V}_{Ü} c_{BR}} = \text{mean residence time of bacteria or sludge age}. \tag{60}$$

By neglecting the bacteria concentration in the overflow of the sedimentation tank it follows from a biomass balance on the sedimentation tank

$$(\dot{V}_o + \dot{V}_R)c_{Ba} = \dot{V}_R c_{BR} + \dot{V}_{ü} c_{BR} \tag{61}$$

or considering Eqs. (56), (57), (59) and (60):

$$t_{vB} = \frac{t_v}{1 + n_R - n_E n_R}. \tag{62}$$

By introducing Eq. (62) into (58) we obtain

$$c_{Sa} = \frac{K_S k_d t_{vB} + K_S}{t_{vB}(\mu_{max} - k_d) - 1} \tag{63}$$

and conclude: c_{Sa} depends only on the sludge age t_{vB} and the kinetic coefficients. For $c_{Sa} = c_{So}$ the critical sludge age follows

$$t_{vBK} = \frac{K_S + c_{So}}{c_{So}(\mu_{max} - k_d) - K_S k_d}. \tag{64}$$

Only for $t_{vB} > t_{vBK}$ is Nitrosomonas able to grow. For $t_{vB} < t_{vBK}$, all nitrifiers were washed out!

From Tables 3 and 4, the following kinetic coefficients can be obtained

$$\mu_{max, 10°c} = 0.29 \text{ d}^{-1}, K_s = 0.7 \text{ mg l}^{-1} NH_4 - N,$$

$$k_d = 0.048 \text{ d}^{-1}.$$

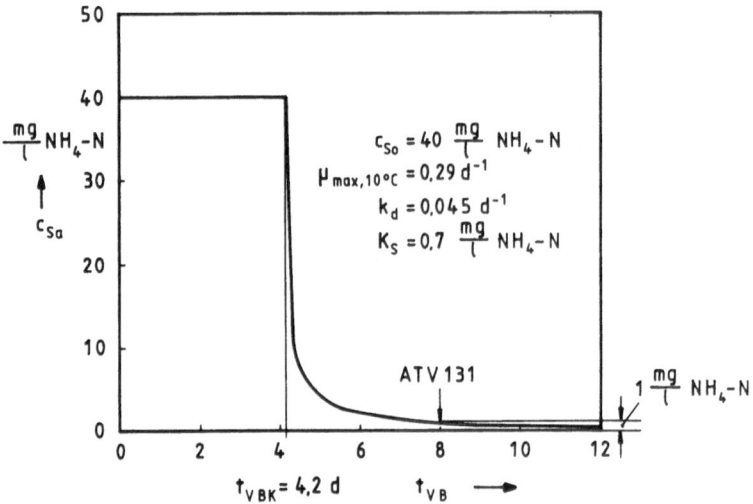

Fig. 8. Ammonia ion concentration in the effluent of a CSTR nitrification system as a function of sludge age (Eq. (63))

For sewage ($c_{So} = 40 \, mg \, l^{-1} \, NH_4 - N$) a critical sludge age of

$$t_{vBK} = \frac{0.7 + 40}{40(0.29 - 0.048) - 0.7 \times 0.048} = 4.2 \, d$$

can be calculated. Figure 8 shows Eq. (63) for the given kinetic coefficients.

Following the guidelines A 131 from the ATV (1991) [45] a sludge age of $t_{vB} = 8 \, d$ must be taken as a basis for the calculation of the reactor volume for plants with $> 100\,000$ population equivalents. From the model an effluent concentration of $c_{Sa} = 1 \, mg \, l^{-1} \, NH_4 - N$ is obtained. With a recycle ratio $n_R = 0.45$ and a thickening ratio $n_E = 3$ a mean hydraulic residence time of

$$t_v = t_{vB}(1 + n_R - n_E n_R) = 19.2 \, h$$

follows.

6 Nitrification of Higher Loaded Wastewater in Lab. Scale Experiments

6.1 Materials and Methods

Two different wastewaters were used (Table 5).

The real wastewater is produced in the sewage treatment plant Marienfelde in Berlin by the dewatering of thermally (50%) and chemically (50%) conditioned sludges.

Table 5. Composition of wastewaters used [17]

Synthetic wastewater 1	Real wastewater 2
For 100 mg l^{-1} $NH_4 - N$:	2.5–5 g l^{-1} DOC
0.472 g $(NH_4)_2SO_4$	7.5–15 g l^{-1} COD
0.05 g K_2HPO_4	5–10 g l^{-1} BOD_5
0.06 g $MgSO_4$ $7H_2O$	1–2.5 g l^{-1} DOC
7 ml of standard solution	as lower fatty acids
3.9 g l^{-1} Na_2MoO_4 $2H_2O$	0.15–0.35 g l^{-1} org N
28.6 g l^{-1} H_3BO_4	0.6–1.2 g l^{-1} $NH_4 - N$
18.1 g l^{-1} $MnCl_2$ $4H_2O$	0.06 g l^{-1} SO_4^{2-}
2.2 g l^{-1} $ZnSO_4$	0.06–0.1 g l^{-1} $PO_4 - P$
0.8 g l^{-1} $CuSO_4$ $5H_2O$	0.5–0.8 g l^{-1} dm
0.5 g l^{-1} $Co(NO_3)_2$ $6H_2O$	10–30 mmol l^{-1}
in distilled water (1 l)	alkalinity

Table 6. Description of the different experiments

Experiment	Wastewater	Reactor	Process	Solid particles
1	1	Fluidized bed 7.55 l	Nitrification	Pumice stone
2	1	Stirred tank 14 l	Nitrification	
3	1	Stirred tank 7.2 l	Nitrification	$CaCO_3$- powder
4	2	Stirred tank 14 l	Denitrification	$CaCO_3$- powder
		Stirred tank 7.2 l	Aerobic carbon removal	
		Fluidized Bed Loop Reactor 10 l	Nitrification	

The results of four experiments are described in the following sections. The experimental plan is summarized in Table 6. The first experiment was carried out in a fluidized bed reactor (Fig. 9). In experiments 2 and 3 the fluidized bed reactor was replaced with a stirred tank reactor. In experiment 4 real wastewater was treated in three stages (Fig. 10). For nitrification a fluidized-bed loop reactor was used as a third stage. The pH was controlled automatically at pH 7.8.

6.2 Nitrification in a Fluidized Bed Reactor with 100 mg l^{-1} $NH_4 - N$ in the Influent (Experiment 1)

All experiments discussed in Sect. 6 are from Dombrowski [17]. Wastewater 1 with an influent concentration of 100 mg l^{-1} $NH_4 - N$ was treated. Small

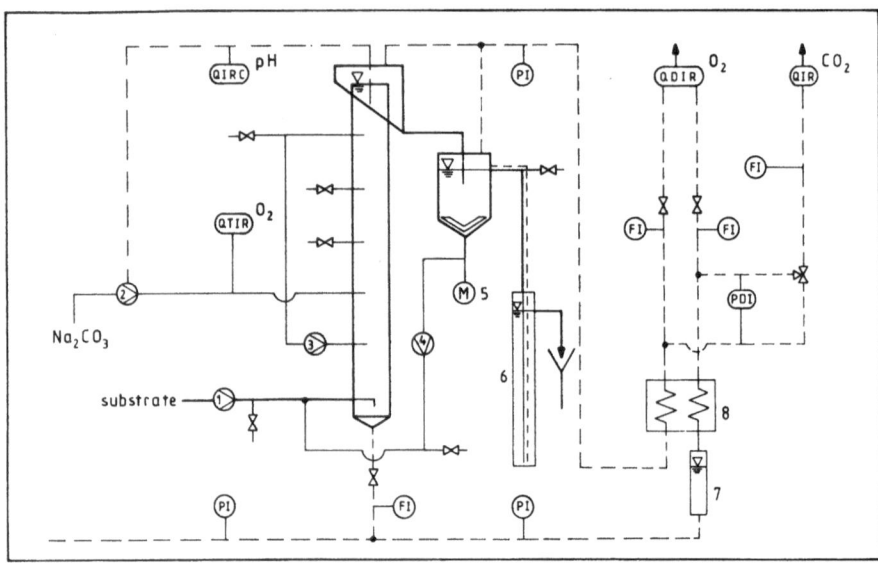

Fig. 9. Experimental set-up with fluidized bed reactor (experiment 1) |17|

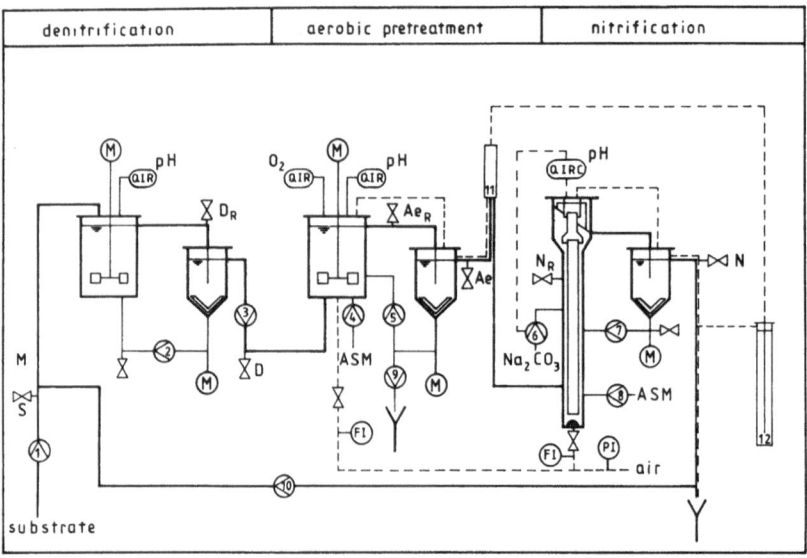

Fig. 10. Experimental set-up of a three stage plant for denitrification, aerobic carbon removal and nitrification |17|

pumice stone particles were used. They were included into the bacterial flocs and weighted them down, resulting in a higher settling rate. A high aeration rate kept the oxygen concentration always $> 6 \text{ mg l}^{-1}$. As seen in Fig. 11a the mean hydraulic residence time could be decreased from $t_v = 14 \text{ h}$ to 1.7 h without

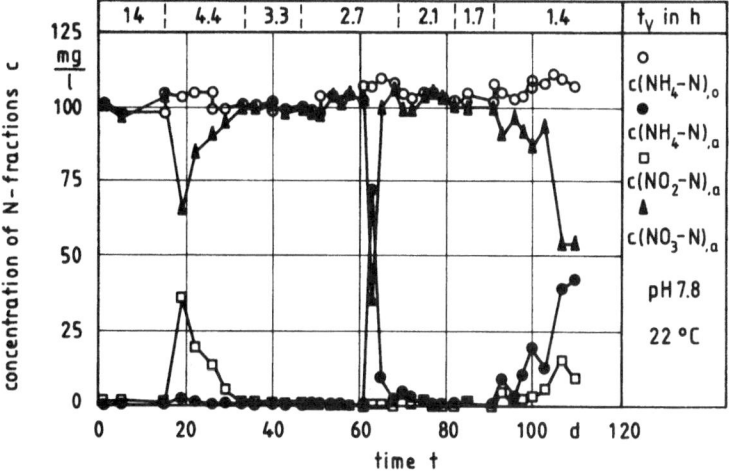

a

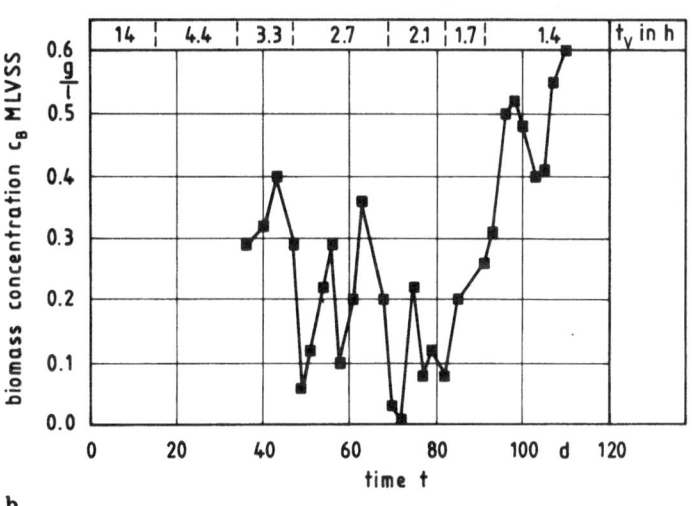

b

Fig. 11a, b. Nitrification in a fluidized bed reactor, pumice stone particles added (experiment 1). **a)** N-fractions. **b)** Biomass concentration in the overflow from protein measurements

increasing the $NH_4 - N$ and $NO_2 - N$ concentrations in the effluent, demonstrating a relatively stable process with almost total nitrification and high reaction rates. An increase in $NH_4 - N$ and $NO_2 - N$ could be observed twice, shortly after t_v was decreased, but they decreased again after some days. However, for $t_v = 1.4$ h the limit is reached, resulting from protein measurements in an increase of $NH_4 - N$ and $NO_2 - N$ concentrations in the effluent (Fig. 11a) and biomass concentration in the overflow of the sedimentation tank (Fig. 11b). Obviously, the mean hydraulic residence time fell below the critical value. This follows from Eq. (64) with $\mu_{max,22°C} = 0.92 \, d^{-1}$, $k_d = 0.045 \, d^{-1}$, $K_s = 0.7 \, mg \, l^{-1} \, NH_4 - N$ and $c_{S_0} = 100 \, mg \, l^{-1} \, NH_4 - N$ to

$$t_{vBK} = 1.15 \, d.$$

From Eq. (62) we obtain the corresponding critical hydraulic mean residence time

$$t_{vK} = 1.65 \, h$$

for a thickening ratio of $n_E = 3$ and a recycle ratio $n_R = 0.47$. From a balance of solids around the sedimentation tank

$$n_{Rmax} = \frac{1}{n_E - 1} \tag{65}$$

follows to

$$n_{Rmax} = 0.5.$$

This process can be described very well with the model from Sect. 5, assuming no oxygen limitation and no substrate inhibition!

6.3 Nitrification in a Stirred Tank Reactor with 300 mg l^{-1} $NH_4 - N$ in the Influent (Experiment 2)

Wastewater 1 with an influent concentration of $300 \, mg \, l^{-1} \, NH_4 - N$ was treated in an aerated stirred tank reactor ($V_R = 14 \, l$) because of the higher oxygen mass transfer rate, needed in order to reduce oxygen limitation. For $t_v = 6 \, h$ a nearly complete nitrification can be observed (Fig. 12a). The reduction of t_v to 4.4 h caused a disturbance in the NH_4-oxidation. Using Eq. (64) we can calculate the critical sludge age

$$t_{vBK} = 1.14 \, d.$$

On the 15th day, when t_v was reduced to 4.4 h, the thickening ratio was only $n_E = 1.33$. Therefore the recycle ratio n_R had to be increased in order to maintain bacteria concentration at the necessary high level. For $n_R = 2.5$ ($n_{Rmax} = 3$) a critical hydraulic mean residence time of

$$t_{vK} = 4.9 \, h$$

follows from Eq. (62), showing that the washing out of Nitrosomonas must begin. Because of the increase of thickening ratio to about $n_E = 2.2 \, t_{vK}$ again decreased, resulting in a slowly decreasing NH_4-concentration, which follows from an enrichment of biomass concentration by Nitrosomonas. However, after the 21st day Nitrobacter was continuously washed out in spite of the later reduction of t_v. Obviously, oxygen limitation has a greater effect on the growth of Nitrobacter, which is in accordance with the different K'-values (Table 3). The model in Sect. 5 must be improved to consider oxygen limitation, if we want to describe the washing out of Nitrobacter!

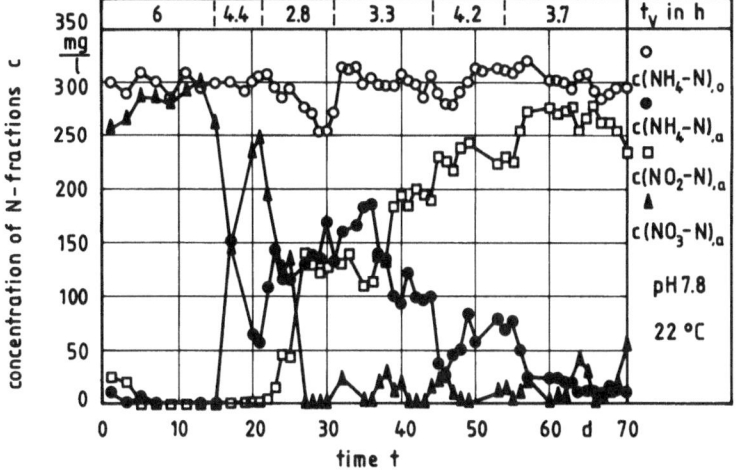

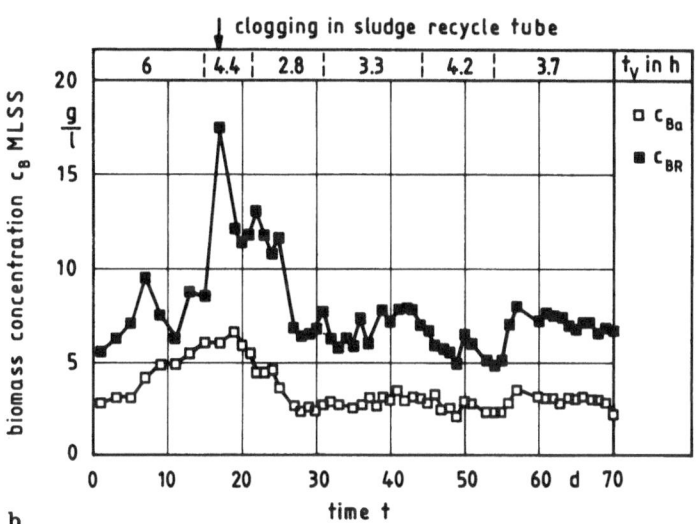

Fig. 12a, b. Nitrification in a stirred tank reactor (experiment 2). **a)** N-fraction. **b)** Biomass concentration in reactor and recycle flow

6.4 Nitrification in a Stirred Tank Reactor
with 1000 mg l^{-1} $NH_4 - N$ in the Influent (Experiment 3)

Wastewater 1 with an influent concentration of 1000 mg l^{-1} $NH_4 - N$ was treated in an aerated stirred tank with a volume of 7.2l. $CaCO_3$-powder was added, in order to improve the settling properties of the formed flocs and to stabilize the pH inside of flocs, which would otherwise decrease as a result of H^+-formation.

During the experiment the hydraulic mean residence time was $t_v = 3.3$ h for a temperature of 25°C (Fig. 13). For the first 15 days pH was maintained at 7.2. The oxidized amount of $NH_4 - N$ was relatively low and no $NO_2 - N$ was oxidized. Between the 16th and the 30th day the pH was increased to 7.8, resulting in a nearly total removal of $NH_4 - N$. Obviously, the growth of Nitrosomonas had been limited by a too-low NH_3-concentration. The increase in pH resulted in a higher NH_3-concentration inside the flocs avoiding substrate limitation. Because of the relatively low K'-value of Nitrosomonas NH_3-oxidation was not reduced considerably by oxygen limitation. However, the growth rate of Nitrobacter was retarded by oxygen limitation and HNO_2-inhibition. The high oxidation rate of NH_3 resulted in a high formation rate of H^+-ions, which was not neutralized totally. Because of the high NO_2^--concentration and low pH, high concentrations of HNO_2 were formed (Fig. 14). Finally, Nitrobacter was washed out as a result of a too-low growth rate. Nevertheless, a high rate

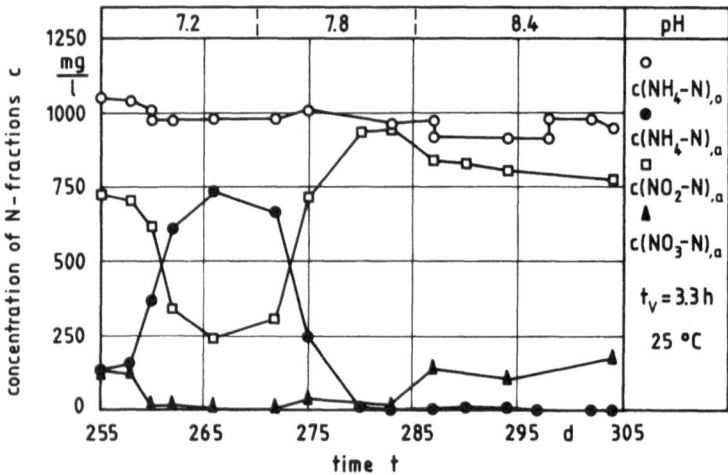

Fig. 13. Nitrification in a stirred tank reactor at $t_v = 3.3$ h at various pH

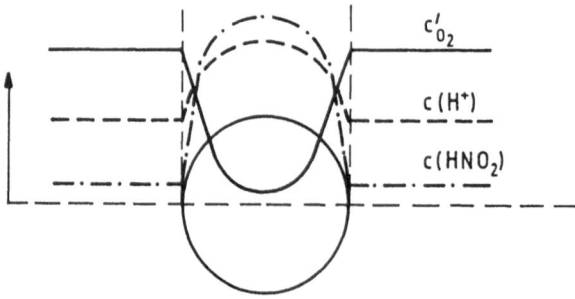

Fig. 14. Qualitative representation of concentration profiles inside a floc for a high rate nitrification process showing unfavourable growth conditions for Nitrobacter

nitrification process without NO_2-oxidation may be important from a practical point of view, because NO_2 can be reduced by denitrifiers.

During this high rate nitrification experiment, the mean bacteria concentration, resulting from protein measurements, was $c_{Ba} = 4$ g odm per l inside the reactor and $c_{Br} = 10.5$ g odm per l in the sludge recycle flow giving a thickening ratio of $n_E = 2.63$ and a maximal recycle ratio of $n_{Rmax} = 0.61$. With $\mu_{max, 25°C} = 1.23\ d^{-1}$, $k_d = 0.045\ d^{-1}$, $K_s = 0.7$ mg l^{-1} $NH_4 - N$ and $c_{So} = 1000$ mg l^{-1} $NH_4 - N$ a critical sludge age of

$$t_{vBK} = 0.84\ d$$

can be calculated from Eq. (64) assuming no oxygen limitation and substrate inhibition of Nitrosomonas. For a recycle ratio of $n_R = 0.5$ a critical hydraulic residence time of

$$t_{vK} = 2.82\ h$$

follows from Eq. (62), showing a theoretical operation limit for the experiments described. Assuming that all Nitrobacter were washed out and that the measured biomass of 4 g odm per l consisted of active Nitrosomonas the specific nitrification rate was

$$\frac{r_{NH_4-N}}{c_{Ba}} = \frac{c(NH_4 - N_0)}{t_v\, c_{Ba}} = 1.82\ g\ NH_4 - N\ g\ odm^{-1}\ d^{-1}.$$

The maximal specific nitrification rate follows from kinetic considerations

$$\frac{r_{NH_4-N,\,max}}{c_{Ba}} = \frac{\mu_{max}}{Y_{B/NH_4-N}} = \frac{1.23}{0.147} = 8.4\ g\ NH_4 - N\ g\ odm^{-1}\ d^{-1}$$

showing, that 22% of the maximal rate could be reached in spite of substrate and oxygen limitation.

6.5 Carbon and Nitrogen Removal from a High Strength Wastewater in a Three Stage Process (Experiment 4)

The composition of the real wastewater 2 is listed in Table 5. Figure 10 shows the experimental set-up. The experiment was divided into three different periods, which can be distinguished by different recycle flow rates from the exit of the nitrification reactor to the denitrification reactor (periods I and II) and the inlet flow rate (periods II and III) (Table 7).

At the beginning of the 70 day experiment and after the 50th day $CaCO_3$-powder was added to the nitrification reactor. The results are illustrated in Figs. 15–17, showing c_s, $c(NH_4 - N)$ and $c(NO_3 - N)$ for five sampling points: influent S, mixing point M, effluent of denitrification system D, effluent of aerobic carbon removal system Ae, effluent of nitrification system N (Fig. 10). Several times fresh wastewater was supplied, which was marked in Figs. 15–17

Table 7. Flowrates and mean residence times for three operation periods, t_{vD}, t_{vAe} and t_{vN} are the real residence times for the three stages, considering the total flow rate $\dot{V}_o + V_R + \dot{V}_u$

Period	$\dot{V}_o$	$\dot{V}_u$	t_{vD}	t_{vAe}	t_{vN}	t_v
	$1\,h^{-1}$	$1\,h^{-1}$	h	h	h	h
I	0.55	4.1	3.2	1.8	2.5	55.5
II	0.55	2.2	5.9	3.3	4.6	55.5
III	1.2	2.8	4.7	2.6	3.5	25.2

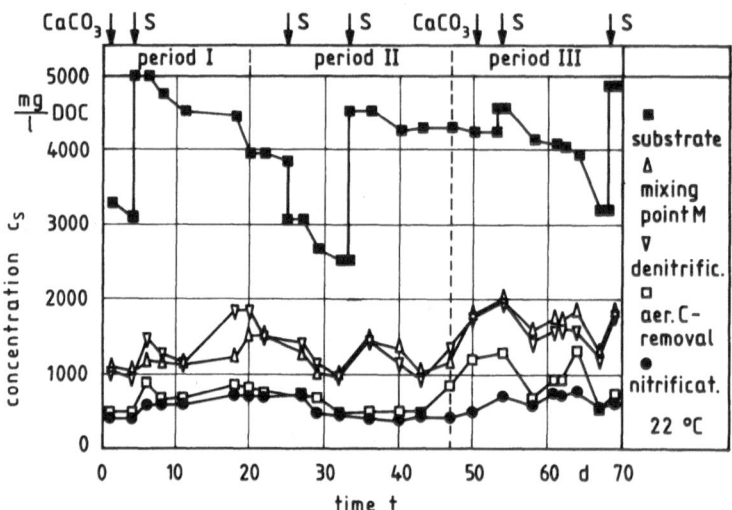

Fig. 15. DOC concentration plotted against time (real wastewater)

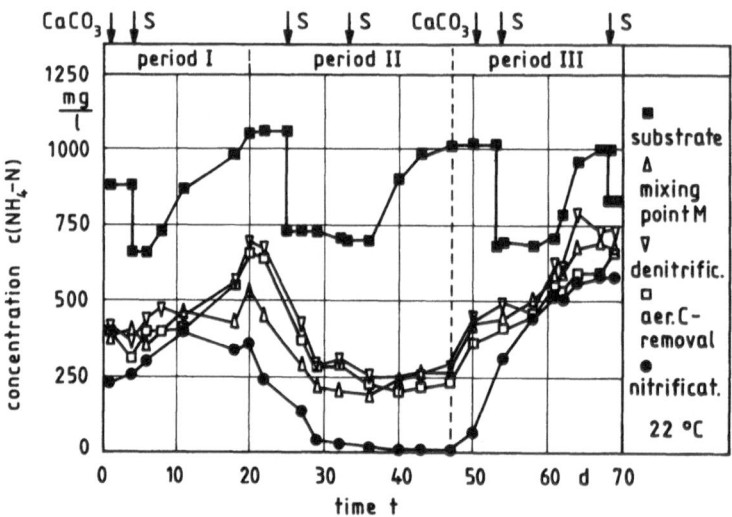

Fig. 16. $NH_4 - N$ concentration plotted against time (real wastewater)

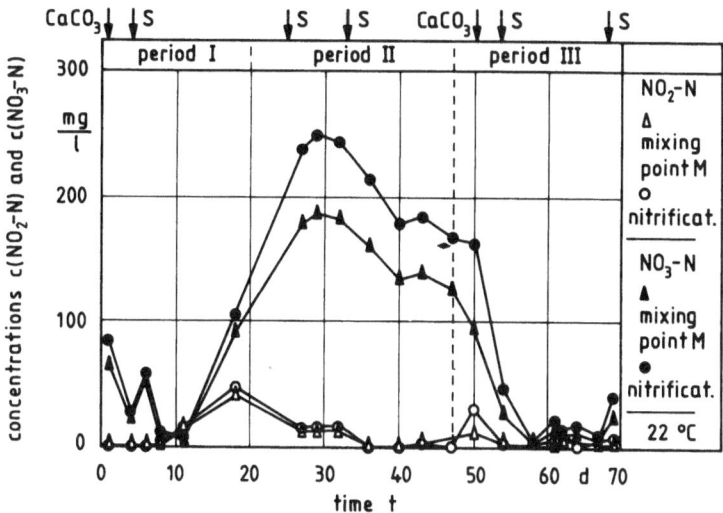

Fig. 17. $NO_3 - N$ and $NO_2 - N$ concentrations plotted against time (real wastewater)

by the symbol ↓. In the storage tank a partial microbial degradation occurred resulting in a decrease in DOC and an increase in ammonia ions by ammonification.

In *period I* carbon was removed from a mean of 4000 mg l^{-1} DOC to 600 mg l^{-1} DOC in the effluent of the nitrification system. The large decrease between the influent and the mixing point is due to the high recycle flow rate $\dot{V}_u$, which was needed to achieve a high degree of denitrification (Fig. 15). However, the nitrification rate was not high enough. Most probably, the concentration of nitrifiers could not be increased because of the high hydraulic load of the nitrification sedimentation tank. This results from the high recycle flow rate and the separation of the sludge into heterotrophs (denitrification and aerobic carbon removal) and autotrophs (nitrification). Figure 17 shows only a low nitrate formation.

In *period II* the recycle flow rate was reduced from 4.1 to 2.2 l h^{-1}, resulting in a lower hydraulic load of the sedimentation tanks. On the 41st day, the sedimentation tank of the denitrification system went out of operation, showing the same results as before: It is not necessary to separate the biomass for the anoxic denitrification and for aerobic carbon removal! Because of an increase in the heterotrophic biomass concentration, the DOC was decreased to 400 mg l^{-1} in the effluent of the nitrification system (Fig. 15). Most of the DOC was reduced in the aerobic carbon removal stage (4300 mg l^{-1} → 400 mg l^{-1} DOC). The remaining substances are not biodegradable!

Because of the reduced hydraulic load of the sedimentation tank, the concentration of nitrifiers could be increased. Therefore, nearly all of the ammonia was oxidized to nitrate (Figs. 15 and 16), which was reduced in the denitrification

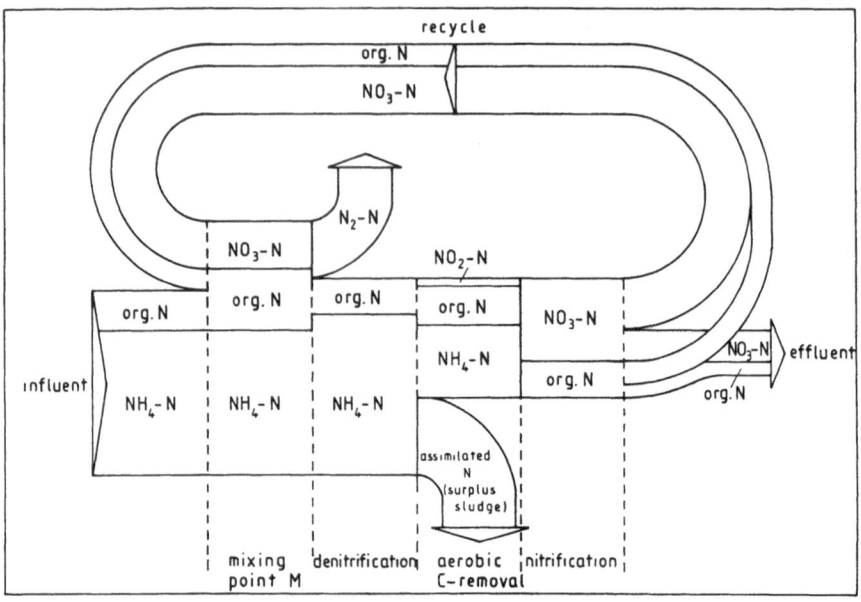

Fig. 18. Nitrogen balance for one day with high nitrification rate (real wastewater)

stage. It was possible to hold back Nitrobacter too, so that almost no nitrite was formed. The total mean residence time, including sedimentation tanks was $t_v = 55$ h. We have to emphasize that this high value results from the high recycle flow rate and the high hydraulic load of the nitrification sedimentation tank.

In *period III* the flow rate was increased from $0.55 \, \text{l h}^{-1}$ to $1.2 \, \text{l h}^{-1}$. But the resulting total mean residence time of 25.2 h was too low for nitrification: a large amount of nitrifiers was washed out (Fig. 16). Comparatively, aerobic DOC removal was not disturbed. If we want to reduce total mean residence time for biological carbon and nitrogen removal from high strength wastewater, it is necessary to immobilize the nitrifiers. Then, it is possible to eliminate the sedimentation tank to hold back the nitrifiers!

Figure 18 shows a total nitrogen balance for one day with a high nitrification rate in a three sludge system. It is remarkable, that a relatively high amount of nitrogen is assimilated by the aerobic heterotrophs!

7 Pilot Plant Nitrification Experiments

Table 8 shows some results of pilot plant nitrification experiments from the last 15 years.

In conventional activated sludge systems for carbon removal and nitrification of sewage the specific nitrification rate of $0.017 \, \text{g} \, \text{NH}_4 - \text{N/g odm:d}$

Table 8. Some results from pilot plant nitrification experiments

Author	Year	Ref.	Wastewater	Nitrification reactor	Reactor volume V_R m³	Flow-rate $\dot{V}_o$ m³ h⁻¹	Mean residence time t_v h	Biomass concentration c_{Ba} g l⁻¹	NH₄ – N influent c_o mg l⁻¹	NH₄ – N effluent c_a mg l⁻¹	Specific reaction rate $\dfrac{c_o - c_a}{c_{Ba} t_v}$ gNH₄ – N/g d
Shade	1977	48	Chemical plant effluent	Activated sludge	2580	48.9	52.8	3.5	200	< 1	0.026
Ganczarcyk	1979	49	Coke plant effluent	Activated sludge, second stage	14.8	0.35	42.0	4.75	460	42	0.05
Pascik, Mann		5	Refinery plant effluent	BAYER Tower-Biology	30	0.73	41.0	3.5	622	140	0.08
Williams et al.	1986	46	Sewage, biologically pretreated	Fluidized bed, sand	30	20	0.50	12	30	1	0.12
Heijnen	1988	50	Yeast plant effluent, anaerobically pretreated	Airlift suspension, sand 0.1–0.3 mm	0.35	0.22	1.60	20	120	4	0.15
Dalentoft	1990	47	Pectin production effluent	Activated sludge denitrification/ nitrification 2 stages			15.4	3.5	450	2	0.20

follows from $c(NH_4 - N)_0 = 40$ mg l^{-1}, $t_v = 19.2$ h and $c_B = 3$ g l^{-1} (Sect. 5). It is relatively low because carbon removal and nitrification occurs in the same reactor with a *one* sludge system resulting in a population mixture of mainly heterotrophs and few autotrophs. In this kind of treatment system, it is not possible to enrich the autotrophic bacteria because the slower growing autotrophs are removed with the surplus sludge. If we want to increase the specific nitrification rate, we have to separate the autotrophic from the heterotrophic biomass. Table 8 shows results for nitrification of biologically pretreated sewage in a fluidized bed reactor [46]. A much higher specific nitrification rate of 0.12 gNH$_4$ − N per g odm per d could be obtained. As a consequence of the high biomass concentration of 12 g odm per l a mean residence time of only 0.5 h was necessary!

The other data presented in Table 8 are results from the treatment of industrial effluents. The relation of organics to ammonia is mostly lower than in sewage so that a higher portion of nitrifiers can be obtained in the sludge of one-sludge systems. In 1977 the specific nitrification rate of 0.026 g NH$_4$ − N per g odm d was considered low. In the following years it was increased gradually by different teams to 0.20 g NH$_4$ − N per g odm h [47] both by aerobic or anaerobic pretreatment and by the improvement of reaction conditions (optimal pH resulting in avoidance of growth limitation or inhibition of Nitrosomonas and Nitrobacter, better aeration resulting in avoidance of oxygen limitation, separation of toxic compounds which avoided inhibition of nitrifiers, better mixing resulting in avoidance of short circuit flow).

Note the maximal specific nitrification rate of $\mu_{max}/Y^o_{B/NH_4-N} = 2.0$ (T = 10 °C) and 8.4 g NH$_4$ − N per g odm per d (T = 25 °C) and the relatively high value of 1.82 g NH$_4$ − N per g odm per d obtained in lab. scale experiments for T = 25 °C (Sect. 6.4). Naturally, such high rates cannot be reached in pilot or full scale plants!

8 Summary

1. Effluents with a high ammonia concentration are produced by many industries. Because of the ecological damage, nitrogen will have to be removed in the future before the stream can be discharged into rivers. In many cases biological methods are economical.
2. For a better understanding of reaction engineering problems arising from nitrification and denitrification, the biokinetic fundamentals have to be studied.
3. The real electron donors for Nitrosomonas is NH$_3$ and for Nitrobacter HNO$_2$. Their concentrations can be changed by pH, resulting in substrate limitations for low concentrations and in substrate inhibition for high concentrations.

4. Oxygen concentration has a more important influence on the nitrification rate than on the aerobic carbon removal rate. Especially for nitrite oxidation it is difficult to avoid oxygen limitation.
5. Reaction engineering models are helpful for a better understanding of the processes and for the designing of reactors and aeration systems. This was demonstrated for nitrification in a conventional sewage treatment plant.
6. In a series of lab. scale experiments the load was gradually increased and the reasons for the limitations in capacity were discussed. These limits may be caused by an excessively high hydraulic load of the sedimentation tank, an oxygen concentration that is too low and a pH inside the formed bacteria flocs which is also too low.
7. In a further lab. scale experiment, the problems arising for carbon and nitrogen removal from high strength wastewater in a three stage process (denitrification, aerobic carbon removal, nitrification) was demonstrated. As a result of the high recycle flow from the overflow of the nitrification clarifiers to the denitrification reactor the sedimentation tanks are highly loaded in two- or three-sludge systems. The process can be improved by using bioreactors with immobilized nitrifiers.
8. Reports of nitrification from industrial effluents in full scale plants are very rare. Pilot scale investigations for the last 15 years show remarkable advances in the increase in nitrification efficiency and in the stabilization of the process. The time is ripe for nitrogen removal from industrial effluents in full scale processes!

9 References

1. Cousins WG, Mindler AB (1972) J WPCF 44:607
2. Koziorowski B, Kucharski J (1972) Industrial waste disposal. Pergamon, Oxford
3. Adams CE, Eckenfelder WW (1977) J WPCF 49:413
4. Garrison WE, Kremer JG, Murk J (1973) Proc 28th Ann Purdue Ind Waste Conf 309–322
5. Pascik I, Mann T (1984) Water Sci Tech Vol 16, Vienna, 215–223
6. Hutton WC, La Rocca Sa (1975) J WPCF 47, 989–997
7. Arnold DW, Wolfram WE (1975) Proc 30th Ann Purdue Ind Waste Conf 760–767
8. Meinck F, Stoof H, Kohlschütter H (1968) Industrieabwässer; Fischer, Stuttgart
9. ATV (eds) (1985) Lehr- und Handbuch der Abwassertechnik, 3 Aufl, Bd V: Organisch verschmutzte Abwässer der Lebensmittelindustrie. Ernst, Berlin
10. Braun R (1982) Biogas-Methangärung organischer Abfallstoffe – Grundlagen und Anwendungsbeispiele. Springer, Wien
11. Neumann H, Viehl K (1966) gwf wasser/abwasser 107, 1151–1154
12. Zall RR (1972) J Milk Food Technol 35, 53–55
13. Basu AK (1975) J WPCF 2184–2190
14. Patterson JW, Minear Ra (1975) State-of-the-Art for Inorganic Chemicals Industry: Commerical Explosives; US EPA 600/2-74-009-b
15. US EPA (Hrsg. 1975) Development Document for Effluent Limitation Guidelines and New Source Performance Standards for the Pressed and Blown Glass Segment of the Glass Manufacturing Point Source Category; US EPA 440/1-75/034-a

16. Brown GE (1975) Land Application of High Nitrogenous Industrial Wastewater; Proc of the National Conf on Management and Disposal of Residues from the Treatment of Ind Waste-waters, Washington DC
17. Dombrowski T (1991) Kinetik der Nitrifikation und Reaktionstechnik der Stickstoff-eliminierung aus hochbelasteten Abwässern; VDI-Fortschrittsberichte, Reihe 15: Umwelttech-nik Nr 87
18. Patterson JW (ed) (1985) Nitrite and Nitrate Nitrogen, in Industrial Wastewater Treatment Technology. Butterworth, Boston
19. Francis CW, Mankin JB (1977) Water Research 11 : 289
20. Jewell WJ, Cummings RJ (1975) J of WPCF 47 : 2281
21. Bode H (1985) Beitrag zur Anaerob-aerob-Behandlung von Industrieabwässern; Veröff des Inst f Siedlungswasser-wirtschaft der Universität Hannover, Heft 64
22. Lompe D (1992) Kinetik und Reaktionstechnik der biologischen Denitrifikation; Dissertation, TU Berlin
23. Frame Wastewater Regulation, Allgemeine Rahmen-Verwaltungs-vorschrift über Mindestan-forderungen an das Einleiten von Abwasser in Gewässer (Rahmen-Abwasser VwV), Anhang 1 "Gemeinden", vom 8. Sept 1989
24. Lohaus J (1990) Korrespondenz Abwasser 37, 660–667
25. Loveless Je, Painter Ha (1968) J gen Microbiol 52, 1–14
26. Haug RT, McCyrty PL (1972) J WPCF 44, 2086–2102
27. Sharma B, Ahlert RC (1977) Water Research 11, 897–925
28. Haldane JBS (1965) Enzymes; Longmans Green London (1930) and MIT Press Cambridge, Mass
29. Knowles G, Downing AL, Barret MJ (1965) J gen Microbiol 38, 263–278
30. Jenkins SH (1969) Nitrification, Water Poll Control, 610–618
31. Gray NF (1989) Biology of Wastewater Treatment; Oxford Science Publishers, Oxford
32. Stankewich MJ jr (1972) Proc 27th Ann Purdue Ind Waste Conf 1–23
33. Wiesmann U (1966) Chem-Ing-tech 58, 464–474
34. Wiesmann U (1989) Wasserkalender, Erich-Schmidt, Berlin, p 117
35. Lompe D, Wiesmann U (1991) Chem-Ing-Tech 63, 692–699
36. Anthonisen AC, Loehr RC (1976) J WPCF, 835–852
37. Bergeron P (1978) Karlsruher Berichte zur Ingenieurbiologie, H 12
38. Nyhuis G (1985) Veröffentlichungen des Instituts für Siedlungswasserwirtschaft und Abfalltech-nik der TU Hannover, Heft 61
39. Richardson M (1985) Royal Society of Chemistry, London
40. Bédard C, Knowles R (1989) Microbiol Rev 53, 68–84
41. La Motta E (1979) J Env Eng div, 655–673
42. Tanaka H, Dunn IJ (1982) Biotech and Bioeng 24, 669–689
43. Strand SE, McDonell AJ (1985) Water Research 19, 345–352
44. Larsen-Vefring W (1992) Simulation der Nitrifikation und anderer bakterieller Stoffwandlungen im Biofilm; Dissertation, TU Berlin
45. ATV-Regelwerk, Arbeitsblatt A 131, Feb 91: Bemessung von einstufigen Belebungsanlagen ab 5000 Einwohnerwerten
46. Williams SC, Harrington DW, Cooper PF (1986) Wat Pol Control 85, 81–89
47. Dalentoft E: Biological Treatment of a high strength nitrogen wastewater, Environ Biotechnol-ogy, Int Symp 22/25.04.1991, Ostende
48. Shade HJ (1977) Chem Eng Progr 73, 45–50
49. Ganczarczyk JJ (1979) Water Research 13, 337–342
50. Heijnen JJ (1989) Large Scale Anaerobic-aerobic Treatment of Complex Industrial Waste Water Using Immobilized Biomass in Fluidized Bed and Air-Lift Suspensions Reactors: GVC reprints Verfahrenstechnik der mechanischen, thermischen und biologischen Abwasserreinigung, Bd 2

Author Index Volume 51

Author Index Vols. 1-50 see Vol. 50

Subject Index